LARS GEJL

WADERS OF EUROPE

BLOOMSBURY
LONDON · NEW DELHI · NEW YORK · SYDNEY

Christopher Helm
An imprint of Bloomsbury Publishing Plc

50 Bedford Square	1385 Broadway
London	New York
WC1B 3DP	NY 10018
UK	USA

www.bloomsbury.com

BLOOMSBURY and CHRISTOPHER HELM are trademarks of Bloomsbury Publishing Plc

First published by Gyldendal, Denmark. Published in the United Kingdom in 2016 by Bloomsbury Publishing.

© Gyldendal A/S, Denmark 2015 translated by Peter Sunesen

Lars Gejl has asserted his right under the Copyright, Designs and Patents Act, 1988,
to be identified as Author of this work.

British Library Cataloguing-in-Publication Data
A catalogue record for this book is available from the British Library.

ISBN: HB: 978-1-4729-4705-5
         ePDF: 978-1-4729-4707-9
         ePub: 978-1-4729-4706-2

2 4 6 8 10 9 7 5 3 1

Front cover (main): Pied Avocets by Aurélien Audevard.
Front cover (top), left to right: Dunlin by Gary Thoburn; Eurasian Golden Plover by Rebecca Nason; Red Knot
    by Aurélien Audevard; Eurasian Oystercatcher by Mya Bambrick / myathebirder.blogspot.co.uk; Stone
    Curlew by Porojinicu Stelian/Shutterstock.
Back cover (left to right): Common Snipe by Rebecca Nason; Curlew Sandpiper by Aurélien Audevard;
Common Sandpiper by Adrian Drummond-Hill; Black-tailed Godwit by Rebecca Nason; Collared Pratincole by
Aurélien Audevard.
Spine: Red-necked Phalarope by Hugh Harrop.
Title page: Juvenile Sanderling. 27.9. LG.

Printed and bound in Livonia Print

To find out more about our authors and books visit www.bloomsbury.com.
Here you will find extracts, author interviews, details of forthcoming events
and the option to sign up for our newsletters.

▶ Green Sandpiper, *Tringa ochropus*.
Adult breeding at breeding ground. 17.4. LG

Contents

Spotted Redshank in silhouette at night roost. 19.8. LG.

PREFACE

Did you know that the North American subspecies of Bar-tailed Godwit undertakes the longest non-stop migration of any living animal? That you can see Ruddy Turnstones at almost all ice free coasts around the world during the winter? That Common Snipes 'sing' using their tail feathers, while brightly coloured female phalaropes court their drabber mates with song? These are just a few of the remarkable habits of the waders, a group of birds notable for their great variety in form and behaviour. Waders form a spectacular and entertaining part of our inland birdlife during the summer season, and along coasts and on wetlands at other times of year.

A wader is easily recognisable by its drop-shaped body with a relatively long bill and long legs. Many of the species display unique features such as the upturned bill of the Avocet, and the elaborate head ornamentation of the male Ruff in spring.

With only 44 breeding wader species in Europe, along with 38 rare vagrants from North America and Asia, the challenge of watching and identifying waders in our region seems manageable. But that's not the whole truth. Many birdwatchers actually ignore waders completely, because several of the smaller species look very much alike and because the plumages vary so much, both individually and by season.

It takes a competent birdwatcher, a good hand-book and many hours in the field to become familiar with the various species, particularly the many *Calidris* sandpipers, which occur in five quite distinct plumages throughout the year as well as a further six intermediate stages.

The aim of this comprehensive book, which both describes and illustrates the waders of Europe and tells their natural history, is to provide the birdwatcher with in-depth knowledge and the tools for confident identification. Moreover, it should provide some insight into the fascinating lives of these birds. It is a privilege to gain insight into their miraculous world, and a pleasure to be able to identify them correctly and to know that the particular species focused in your binoculars right now is on its way to the far south of Africa, or its breeding grounds in the Siberian tundra.

Enjoy!
Lars Gejl.

Bar-tailed Godwits, Red Knots and Dunlins gather to feed at the Wadden Sea. 4.5. LG.

ACKNOWLEDGEMENTS

Photographing waders is a tall order indeed. Suitable gear is often heavy and costly and the individual photographer needs to build up all-round knowledge and field experience to find the birds, which are often only easy to observe for a few weeks of the year. Also when the entire undertaking goes down the drain because the weather changes and the birds have flown, an infinite degree of patience is required. The photos in this book have been gathered primarily from a handful of enthusiastic Nordic nature photographers who know their gear and the birds intimately, and who are sufficiently fanatical to travel round the world, wander for days in the mountains, sit tight in hides for hours and crawl on their bellies across mudflats in order to get the perfect picture at just the right time and place. A clear photo taken in the right light conditions and with supplementary background is a pictorial representation which is not only pleasing to look at, but often provides information about the behaviour and environment of the bird and is useful for field identification, showing key plumage details. It also captures the sheer pleasure provided by waders – their colours, shapes and variety. Supplementary pictures have been provided by knowledgeable and skilled photographers from four continents, to extend the scope of the book with rare species and the plumages seldom seen during the European winter. So, heartfelt gratitude is due to the photographers whose work appears in this book and their zeal and commitment to this project.

Special thanks to the following:
Kevin Karlson for photos of North American species and for updates on new identification features distinguishing between Short-billed and Long-billed Dowitchers as well between American and Pacific Golden Plovers.

To Peter Sunesen, taxidermist and field ornithologist, for constructive discussions about plumages and age determination as well as professional criticism.

And not least to Axel Kielland, our publisher at Gyldendal, for believing in, supporting and encouraging the project all the way.

Moreover, it is certainly appropriate to extend gratitude to the many field ornithologists, authors and wader enthusiasts whose efforts and special knowledge have provided help and inspiration. See the lists of literature and information sources at the back of the book.

The author in the field between Rømø and Sylt in the National Park, Wadden Sea. The Wadden Sea extends in its entirety from Ho Bay at Esbjerg, along the German coast to Den Helder in Holland, and it constitutes the world's largest continuous tidal area. It is of international significance for 12 million waders and water birds from large parts of the Northern Hemisphere. The birds are dependent on this nutritious ecosystem, using it as a place to feed, breed, visit on migration and spend the winter.
Photo: Axel Kielland.

The shorebird order

The order *Charadriiformes* numbers some 379 species. It includes families such as hemipodes, gulls and auks, as well as the waders.

The true waders, characterised by a sleek, drop-shaped body, long legs, and in most a medium-long to long bill of variable shape, are represented by up to 214 species.

In Europe 44 species breed regularly and a further 38 species occur as vagrants from Africa, North America and Asia.

The wader's needs

Like many other migrant birds, over thousands of years, waders have sought, found and adapted to the habitats which provide them with optimal life conditions in relation to their requirements throughout the year; in particular, their need to find enough food to fuel their long and hazardous migrations. In the temperate and tropical regions where most species of waders spend the greater part of the year, the supply of food is consistent, and the expenditure of energy is relatively low due to the mild climate.

What waders need in order to raise one or several broods of young in summer is a surplus of protein-rich food, which is accessible without the

▶▶ The map opposite shows the main routes of migration used by, for example, godwits and *Calidris* sandpipers which each spring and autumn migrate to and from Europe, particularly Fennoscandia, and the Arctic to breed.

One of the migration routes, the East Atlantic Flyway, stretches from the Cape of Good Hope in South Africa, up along the eastern Atlantic to northern Europe, where it branches out towards North America and Asia.

The other route, the Mediterranean Flyway, runs from the inner parts of the African continent, over the Mediterranean and eastern Europe towards the Russian and Siberian breeding grounds.

These routes are just two of ten global flyways linking birds' winter quarters in the Southern Hemisphere with the breeding grounds in the Northern Hemisphere.

▲ This Red Knot is showing how the outermost tip of the bill is flexible and can be used as forceps. The tip of the bill is packed with sensitive pressure sensors connected with nerve endings, so that in a few hundredths of a second and at a distance of a few centimetres the bird can decide whether the resistance it feels through its bill when probing into mud, sand and water is produced by a stone, or a potential food item.

Most waders that feed primarily by probing have medium-long or long bills, which are adapted to finding food such as shellfish and worms in a soft substrate. 5.8. JL.

▼ The individual families have developed different bill shapes and methods of feeding so that every nook and cranny of the terrain can be explored, and unnecessary competition can be avoided.

The Little Ringed Plover primarily uses its eyesight to find prey, and captures it with a running and snatching technique.

Like the other plovers it employs the 'paddling' or 'pumping' method on damp soil, as here, where the bird is stamping with one foot, producing vibrations in the soil that simulate the effect of heavy rain. This encourages worms to come to the surface. 4.8. LG.

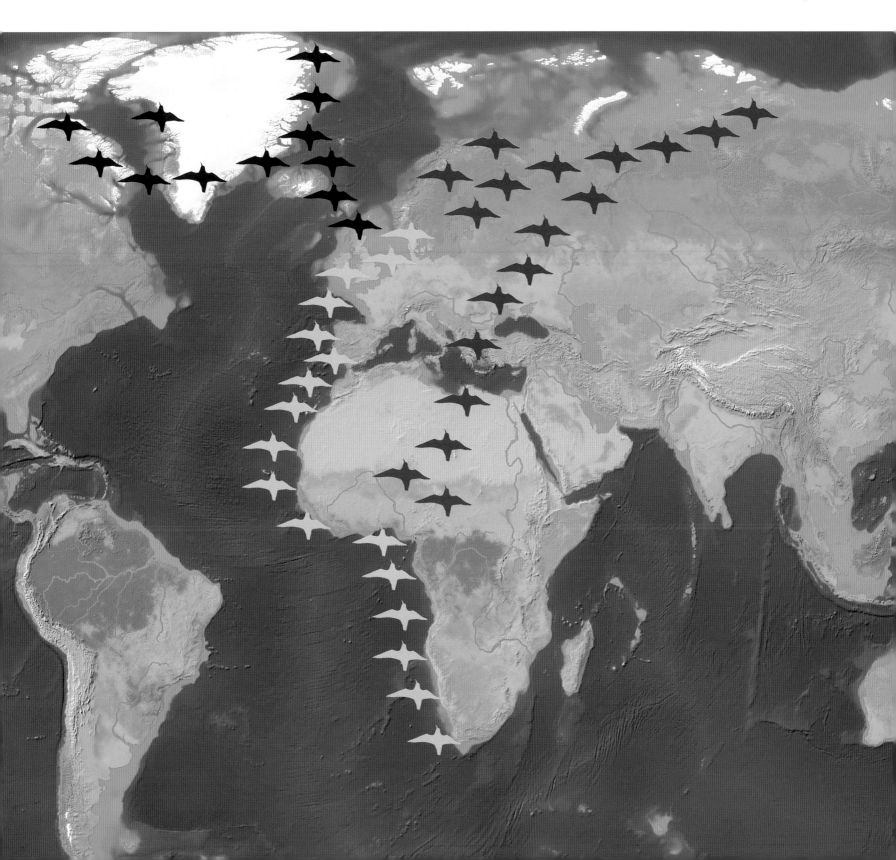

competition that prevails in the winter quarters. In winter, hundreds of thousands of waders can gather at relatively narrow but very food-rich shorelines, with a hinterland consisting of desert, mangrove swamp or rainforest – no good for foraging. However, the daily rise and fall of the tide, and the exposed nature of this habitat, makes these food-rich shorelines generally unsuitable for nesting.

Accordingly, waders migrate long distances and move inland in order to find nesting grounds and to feast on the myriad insects that multiply explosively every summer in the wetlands of the Arctic taiga and tundra.

Northward Ho!

From the two primary wintering areas along the west coast of Africa, the Bijago Archipelagos off Guinea-Bissau and Banc d'Arguin off Mauritania, where millions of waders spend the winter months, the first leg of migration towards the breeding grounds begins in April.

Red Knots and Bar-tailed Godwits divide the long journey into two stages. Other species, such as the *Tringa* sandpipers which winter in continental Africa, split the journey into several shorter and less energy-demanding stages, crossing land where it is easier for them to find food.

For the flocks of godwits and those species that follow the coastline along Africa towards Europe and onward to the Arctic, the first goal on the route is the Wadden Sea along the Dutch, German and Danish western coasts.

Like many other wetlands on migration routes, the Wadden Sea is of vital and international importance for the water birds and waders that use the

From mid March the Wadden Sea is overcrowded. Sandpipers and godwits pour down from the sky, forming assemblies of thousands. At ebb tide they forage on the tidal flats to build up their energy reserves before resuming their journey. 20.4. L.G.

area as a resting, feeding and wintering ground. The vast concentrations of birds that visit the area during spring and autumn provide an amazing spectacle for birdwatchers.

Pitstop

In the Wadden Sea, the increasing daylight period of spring kick-starts the production of filamentous algae, which by virtue of the tidal forces flows over the muddy and sandy tidal flats twice a day. This algae forms the basis of the cornucopia of food that nourishes everything from the infinitesimal mud snails, worms and crustaceans, to fish, birds and seals.

Migrating birds make the most of this opportunity and enjoy a well-deserved rest on the food-packed tidal flats for up two months in order to fill up energy reserves that have been exhausted during the first 4,000 kilometres of the journey from western Africa to Siberia.

In the Wadden Sea, northbound waders spend their time eating and moulting the worn, often grey winter plumage, which is replaced by a smart, often conspicuously bright breeding plumage: the fresh plumage is a signal of good condition, which makes them attractive to potential mates.

The enormous intake of calories during the stop-off allows the waders to store plenty of fat as fuel vital for the onward migration. Each bird may increase in body weight by up to 50%. The fat is primarily stored on the back, breast and belly. With experience, this obesity can even be seen on live birds in the field.

▲▶ Bar-tailed Godwits and Red Knots, some beginning to transition from non-breeding to breeding plumage. 14.4. L.G.

▶ Bar-tailed Godwits, adults in breeding plumage. In just 20 days the Bar-tailed Godwit radically changes its appearance by moulting the pale grey body feathers to a conspicuous, brick-red breeding plumage. Males are smaller than females, and more brightly coloured. 4.5. L.G.

Ready, steady, go!

The departure from the Wadden Sea is controlled by an inherited, genetic programme which triggers off the different stages of the bird's annual cycle. Fuelling has now taken place in the form of maximum stored fat, while excess weight is minimised by shrinking organs such as the liver, stomach and intestines and also leg muscles, which are not heavily used during migration. The sexual organs, which have been dormant during winter, do not develop fully until the bird reaches the breeding area. The body feathers have been moulted and the fresh plumage provides improved insulation. Its colours match the terrain of the breeding area, thus providing excellent camouflage.

Fair winds

Typically, the birds leave their resting areas on a clear evening just after sunset when, through sensors in the middle ear, they sense that the air pressure is rising, promising fair winds for migration.

Parties of 30–40 Bar-tailed Godwits rise, calling nervously and ascending quickly while sorting out their migration flight formation, which is either band- or wedge-shaped.

Once underway, migration often takes place at very high altitudes. The party may even rise to a height of six kilometres to find the tail-wind that increases their own speed of a good 50–60 kilometres per hour up to 70–80 kilometres per hour, perhaps even more.

The birds are bound for the tundra in western and central northern Siberia, a further 4,000 kilometres on from the Wadden Sea. The distance is covered in a non-stop flight lasting two and a half days. While travelling, they hold a steady course by navigating by the sun, the stars, the Earth's magnetic field and, in cloudy weather, ultraviolet bands of polarised sunlight. Additionally, the less 'high-tech' senses such as eyesight and sense of smell are

A mixed party of Bar-tailed Godwits and Red Knots, nearly all in full breeding plumage, are on the point of falling into a typical wedge- or V-shaped migration formation, in which the birds at the rear are sheltered in the slipstream of the ones in front and are practically pulled along.

This technique, which is also employed by team cyclists, permits the birds at the rear to save energy while the ones in front break the wind resistance and create a shelter.

During migration the birds regularly switch positions so that each in turn benefits from the trick. 5.5. LG.

employed when the smell of the tundra is detected and the recognisable pattern of rivers and pools comes into view. Should the wind become too powerful and the party is in danger of being forced off course, they descend towards calmer wind conditions. The tail-wind when migrating helps the party to conserve energy and thereby minimise loss of the precious fat reserves. If the migration takes place according to plan without unexpected low pressures, head-winds or blizzards, and if their arrival at the breeding grounds has been so well timed that the ice has melted and the pools are free of ice, the birds still have a surplus of fuel and are well set to begin the breeding process right away.

Fun and games

The Bar-tailed Godwit is monogamous. The male starts off the breeding period with vocal display flights over his territory.

Practically all wader species proclaim their territories by using song combined with an aerial display, which can vary from somersaults, whirring flight, slow paddling flight and swift racing low over the ground, to steep dives from a great height with mechanically generated sound, such as the 'drumming' (made by tail feather vibration) of the Common Snipe.

Performing display flights demonstrates strength and ample energy, to impress potential mates and to discourage rivals.

However, it's not always the male who's in charge during courtship behaviour among waders, nor are conventional pair bonds always formed. In the Red-necked Phalarope and Eurasian Dotterel, the more brightly coloured female performs the display flight and defends the territory. With Ruffs, which are practically silent throughout the year, females seek out the silently dancing males on an arena, and select a partner from the throng. And some male Pectoral Sandpipers become nomadic and fly across the entire breeding area from western Siberia to Alaska, in order to mate with available females.

Eggs, chicks and responsibility

If the female is in good shape, she begins to lay a clutch of eggs (2–5 but typically four) within a week of arrival on the breeding grounds (with an interval of one or two days between laying each egg).

The eggs are surprisingly large and require a comparatively long incubation of 20–25 days. Because the wader chicks hatch in a well-developed state, they can follow the parent bird away from

▶ In open habitats, nests with begging young in need of continual feeding would be far too conspicuous. Consequently the waders' chicks are ready to face the world just a few hours after hatching.

The two newly hatched chicks in the nest are decorated with a sprinkling of 'powder puff' down, as the pale tips of the chicks' first downy feathers are called.

The powder puff down gives the plumage a broken pattern, making the chicks practically invisible.

A few hours after hatching, when the chicks are dry, they follow the parent bird away from the nest and immediately catch and consume their first insects. 15.7. Nunavat, Canada. JKAM.

▲ In the tundra the mixture of dark colours demands other types of camouflage to the uniformly coloured light sand flats of the winter quarter. The deep brick-red colour of the underside, in combination with the golden and black stripes of the upperside, makes this incubating male Bar-tailed Godwit almost indistinguishable from his surroundings. 1.6. Yukon, Alaska. JKAM.

the nest, often to a better feeding area, immediately after hatching and drying.

Bar-tailed Godwits are among the few waders where both adults share the parental duties. In the case of the Temminck's Stint, the males copulate with several females and vice versa. The female lays a clutch fathered by her first mate, which he incubates and looks after. Next, she pairs with another male and lays a second clutch, which he incubates, and later may produce a third clutch which she incubates on her own.

In contrast, the Curlew Sandpiper male leaves the breeding ground as soon as the female starts to incubate, whereas in the Spotted Redshank and several other species it is the female that dashes off, leaving the male to incubate and rear the young.

The various breeding strategies ensure that the next generation receives the strongest genes and best care, therefore the highest chances of survival.

The reason why the male is often left alone with the clutch is that generally the female is larger, and would thus present the chicks with more significant competition for food. The opposite is seen in species with larger males, such as Ruff.

Food and danger

The timing of hatching coincides with the peak emergence of insects and other invertebrates; a brief period during the relatively cool and short summer. Protein-rich insects make up the main part of the diet for small chicks.

As already mentioned, wader chicks are capable of finding food for themselves a few hours after hatching. Only a few species, including Eurasian Woodcock, Eurasian Oystercatcher and the snipes, feed their offspring.

Most chicks simply follow the parents and pick up winged insects and other small creatures until they learn how to use their growing bills as probes and begin to feed like the adults on worms and other items in the mud.

Depending on the species, the young waders are capable of flying after 20–30 days, but survival rate is rather low because of predation, adverse weather

Over many millennia, birds have developed plumage which provides each species with all that it needs in terms of insulation, protection and camouflage, in the best possible way.

The birds also know instinctively where best to place the nest and and where they themselves should rest and sleep so as to obtain the maximum benefit of the patterns and colours of their feathers.

Here, it seems that this juvenile Little Stint is conscious of the best place to rest, in order to be less noticeable in its surroundings.
6.9. Sweden. DP.

conditions and fluctuations in insect supplies. In years with low numbers of lemmings on the tundra, gulls, skuas, owls, birds of prey, foxes and stoats take a heavier toll on the flightless young waders, and even adult birds.

When a predator approaches, the parent bird often simulates injury or illness by fluttering away from the young, calling and trailing a seemingly broken wing, hoping thereby to divert the predator. Mostly these 'distraction displays' are successful, but some experienced Arctic Foxes correctly interpret the trick as evidence of young birds nearby.

If a whole clutch of eggs or brood of young is lost, the parent birds may nest again, but if the physical condition of the female is less than perfect, further breeding attempts are abandoned and the adult birds quickly seek out feeding areas where they can rebuild the necessary resources for the autumn migration.

Back again

For the birds that have fulfilled their parental duties, it is time to return to the winter quarters.

Some species such as the Great Snipe migrate directly to the wintering grounds in tropical Africa in a matter of a few days. Several other species follow the shorelines in several shorter flights, and the first adult birds (some of which are free to travel early because they failed to breed successfully) can be seen returning in June, after a mere two or three weeks' stay in the breeding territory.

At habitual roosting areas such as the Wadden Sea, they once more build up fat reserves while moulting the worn breeding plumage, including the flight feathers. The fresh winter plumage is lighter-coloured than breeding plumage, again matching the main colours of the winter habitat.

Later in summer the juveniles arrive, occasionally guided by adult birds but mostly unaccompanied and relying on their inherited migra-

tory instincts. Later in autumn the large flocks leave the shallow waters, which no longer produce a sufficient surplus of food owing to shorter daylight hours and reduced temperatures.

On clear autumn evenings early in October, parties of Bar-tailed Godwits – this time consisting exclusively of juveniles – take off bound for western Africa. With their departure the evidence of the hectic Arctic summer vanishes. Left behind on the beach lie moulted feathers, and only the hardier species like Eurasian Oystercatcher and small groups of Dunlins add a little life to the quiet mudflats.

Even though waders are swift and agile in their flight, they may still fall victim to predators.

In the mangroves along the shores of western Africa lurk mongooses, birds of prey and other dangers. On the breeding grounds there are stoats, foxes, ravens, falcons and even Polar Bears. And during the autumn migration the local birds of prey take a heavy toll on any inexperienced and tired birds that happen to stray away from the main party.

Nevertheless, in spite of poor survival rates among young birds, many waders live surprisingly long lives.

A ringed Eurasian Oystercatcher holds the unofficial world record, reaching the age of 43 years and 4 months. Eurasian Sparrowhawk with Sanderling. 20.8. JL.

Taxonomy

Taxonomy is the scientific discipline concerned with the classification of all living things in a nested hierarchy, according to their relatedness to other organisms. The first level of division is kingdom, followed by phylum, class, order, family, genus and species. Waders fall into the animal kingdom, the phylum Chordata (animals with spinal cords – mainly vertebrates), the class Aves (birds), and the order Charadriiformes (which includes all wader families but also other families such as gulls and auks).

Genetic research has contributed and still continues to contribute a great deal of new knowledge about the mutual relationships among birds, allowing us to refine the accuracy of our classification system. One consequence of this ongoing discovery is that scientific names sometimes have to change.

In the accounts of the individual species of this book, the birds are described broadly in taxonomic order, but at certain points this has been adjusted a little in order to provide a better basis for comparison.

Representative species

In this book the Dunlin, *Calidris alpina*, has been chosen as representative of all small and medium-sized waders.

The Dunlin is the most common and most often observed wader in Europe. The birdwatcher will benefit from becoming familiar with the silhouette, plumage, jizz, voice and choice of habitat of this key species.

This will help establish order in the chaos, particularly for the beginner and make it easier to distinguish, for instance, between the most similar-looking *Calidris* species, but also between *Calidris* species and, for example, some of the *Tringa* species.

Common Greenshank, post juvenile, showing typically serrated *Tringa* tertials and a lower row of new, grey scapulars. 9.8. LG

English and scientific names

Latin and Greek form the basis of the scientific language which is used across frontiers and language barriers to name living things, so that everybody can refer to the same species without the risk of misunderstanding.

In the species accounts in this book, the English name is given first and is followed by the scientific name of the bird in question. The first word in the scientific name is that of the bird's genus, and the second is the name of the species itself.

Meaning of the name

The scientific name of each species is explained in terms of its origin or meaning from the Latin or Greek.

Jizz

The bird's measurements for length and wingspan is given in centimetres.

L = the length from tip of the bill to the tip of the longest tail feather. In species where the bill is particularly long, the measurement of the bill alone is also given.

Ws = wingspan; the distance between the tips of the wings when fully spread.

The general impression of size and shape, along with style of movement, both on the ground and in flight, colloquially termed jizz, is a concept which refers to the overall appearance and manner of a bird.

The physical characteristics are often mingled with an indefinable physiological aspect, with the result that the experienced birdwatcher is able to identify most birds to species, family or at least suborder over long distances, and often without using binoculars.

Similar species

Several species are very similar. Where relevant, this section deals with the most obvious risks for misidentification for each species.

Plumage and identification

Birds' plumage becomes worn and changes its appearance all through the year. Some species look much the same all the year round, but all species moult their feathers regularly, just as we buy new clothes to replace those that have become worn out.

Studies of moulting cycles of individual species are complex but interesting, and are of great value to make reliable identification of species and accurate judgement of age. See more in the section on feathers and moulting.

Plumages are described in detail for different age classes and (where relevant) differences between male and female plumages are also described.

See scientific and age determination as well as moult sequences, page 35.

Subspecies

In cases where a given species is subdivided into geographical subspecies, the other forms are described here and a third name is added to denote subspecies.

The subspecies originally first described within each species is called the nominate, and its third name is always the same as its second.

For instance, the scientific name of the nominate subspecies of Common Ringed Plover is *Charadrius hiaticula hiaticula*. The Arctic and practically identical subspecies has a different name added – *Charadrius hiaticula tundrae*.

Voice

In many situations, knowing the calls and songs of individual species can greatly facilitate identification of birds. And of course listening for calls will help you to locate them in the first place.

However, any attempt to convey bird vocalisations in words will be subjective.

Habitat

This section covers breeding and wintering habitat, and habitats used while on migration.

Breeding biology

Describes courtship, pairing, nesting and care of eggs and young..

Migration

Most of the waders that we observe in Europe are migratory and we can only enjoy them for a few months each year.

Therefore, an understanding of migration periods, migration routes, migration hotspots and winter quarters is important not only for a successful day of birding, but also because it provides a global perspective for your observations.

The following terms are used in the sections on migration.

Each month is divided into three parts:
Early: 1st–10th.
Mid: 11th–20th.
Late: 21st–31st.

Distribution

This section is a brief introduction to the bird's distribution in Britain and Europe, and beyond, as well as its population.

Abbreviations

♂ = male   ♀ = female
juv   juvenile
ad   adult
br   breeding   non-br non-breeding

Photographs

The initials of each photographer are given at the end of each photo caption, and when no locality is stated, the bird has been photographed in Denmark.

Dunlin. Adult breeding en route towards the arctic breeding areas. 5.5. LG.

Technical terms

Arm: the inner part of the wing, from the carpal joint to the body.

Axillaries: the feathers of the armpit.

Base of bill: the part of the bill closest to the face.

Carpal joint: the wing-bend on the front edge, between hand and arm.

Carpal patch: a contrasting patch of colouring at the carpal joint (wing-bend) on the underwing.

Ear-coverts: the feathers covering the ear opening. This area is also known as the cheek and is situated just below and behind the eye.

Eye-ring: narrow circle of short, often pale feathers around the eye.

Eye-stripe: a stripe, usually dark, through or behind the eye.

Fennoscandia: Norway, Sweden, Finland, Russian Karelia and the Kola Peninsula.

Fingers: the spread tips of the outer primaries.

Foot projection: the protruding part of the toes, sometimes also tarsus, beyond the tip of the tail of the flying bird.

Fringe: frayed, pale, often whitish edge to a freshly grown feather, which is quickly worn off.

Habitat: the type of natural environment which best fulfils the feeding and breeding requirements of the species.

Hand: equivalent to outer wing, the part of the wing beyond the carpal joint.

Hybrid: a cross between two different species.

Lore-stripe: narrow, often dark stripe on the lore (the area between the eye and the base of the bill).

Morph: variant of a certain appearance within a species.

Moult: the renewal of the plumage, whereby the worn feathers are shed and new ones grown in their place.

Moustachial stripe: dark, narrow stripe from the base of the lower mandible and along the lower edge of the cheek.

Orbital ring: narrow ring of bare and often brightly coloured skin immediately within the eye-ring.

Overshooting migration: this term is used when birds migrate in the right direction but keep going for a longer distance than normal, and 'overshoot' their goal.

Panel on coverts: contrasting pale area on the upperside of wing.

Polyandry: when a female pairs with more than one male.

Polygyny: when a male pairs with more than one female.

Population: the total number of individuals of a given species within a defined area.

Primary patch: pale patch at the base of the primaries on either upper- or underwing.

Primary projection: the part of the primaries which on the folded wing extend beyond the tip of the tertials.

Reverse migration: mostly occurs in the autumn when for unknown reasons some birds turn the course of their migration by 180 degrees away from the normal direction.

Species: can be defined as a group of individuals with identical traits whose members can interbreed freely, but not with individuals from other similar groups (species).

Supercilium: a stripe, usually pale, above the eye.

Tarsus: the long, visible part of the leg between the toes and the ankle bend (often wrongly called 'knee').

Tertials: the innermost secondaries, most often three, frequently deviating in shape and colour from the rest of the secondaries.

Tibia: the part of the leg above the ankle bend – not easily visible in shorter-legged species.

Vermiculation: fine and wavy barring.

Width of wing: width of the wing at the base.

Window: pale panel of translucent primaries.

Wing coverts: rows of feathers on the hand and arm.

Divided into lesser, median and greater coverts.

Distribution maps are costly to manufacture and expensive to update.

Consequently the distributions of the species are described in a general way and more specifically where it was necessary to provide an overview of the occurrence of the species in the winter and summer seasons.

The map (right) shows the geographical demarcation line between Europe and Asia.

Map 21

Topography

The 'landscape' of birds is divided into logical sections with fixed ornithological terms.

The broader divisions include the main body parts of the bird, while the plumage is divided into particular feather groups or tracts.

On the following three spreads, the feather tracts and the more general parts of the bird are explained to aid field identification.

1. Bill from base to tip.
2. Nostril.
3. Crown.
4. Nape.
5. Mantle. The feather tract between lower nape and back.
6. Upper scapulars. Three rows (along the mantle).
7. Back 'V'. Pale lines on each side of the mantle.
8. Greater scapulars. Two rows.
9. Scapular 'V'. Formed by pale feather edges.
10. Primaries. Primary projection.
11. Tertials. The three innermost secondaries.
12. Greater coverts. One row.
13. Vent.
14. Tibia. The upper visible part of the leg.
15. The heel.
16. Median arm coverts. One row.
17. Tarsus. The lower visible part of the leg.
18. Hind toe.
19. Toes.
20. Belly.
21. Lesser arm coverts. Several rows.
22. Breast.
23. Neck.
24. Ear-coverts.
25. Cheek.
26. Throat.

Juvenile Dunlin, 3.9. LG.

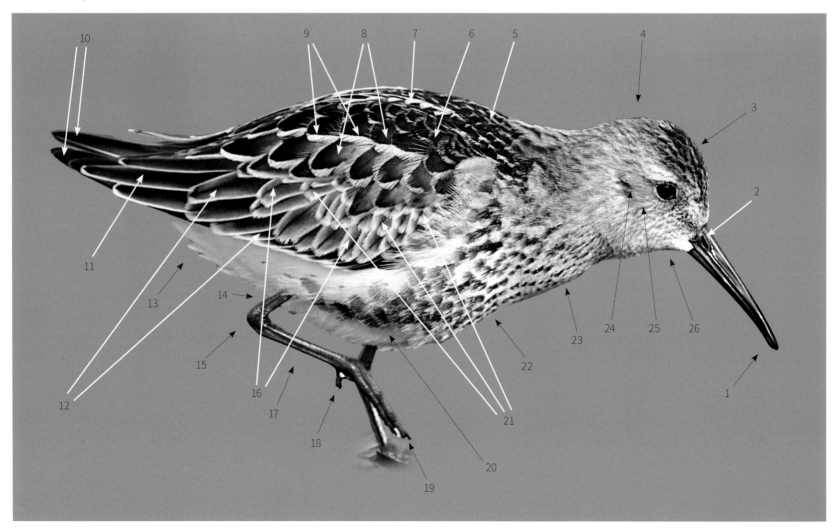

With a resting bird, note:
Length and shape of the bill.
Shape of the head and head markings.
Length and colour of legs.
General colours.
Behaviour and shape.
Mantle.
Scapulars.
Wing coverts.
Length of the tertials, and tertial markings.
Primary projection beyond tertials and tail.

With a flying bird, note:
Length and shape of wings.
Wing-bars.
General markings on the upperwing.
Pattern of mantle and scapulars.
Shape of the tail.
Colour of the tail feathers.
Markings of the uppertail coverts.
Markings on the rump.
Projecting feet or toes beyond the tail.
Contrast between worn and new feathers.

1. Pale back 'V'. Formed by the upper scapulars.
2. Mantle.
3. Back.
4. Outer vane. The part of the feather outward from the shaft.
5. Inner vane. The part of the feather inward from the shaft.
6. Rump. The part between the back and the uppertail coverts.
7. Uppertail coverts.
8. Tail feathers. Variable in number but generally 12.
9. Wing-bar. Varies in width and length. May be double or absent.
10. Broad beige or pale fringes on median and smaller wing coverts often indicate a juvenile.

Juvenile Dunlin. 9.8. EFH.

Upperside–feather groups and details

Juvenile Ruff. 1.9. NLJ.

This upperside image clearly shows the individual feather groups, and plumage details that most waders share.

With brief observations, try to check length and shape of bill, contrast between feather groups, wing-bars, general colours, and tail-patterns; also whether toes or legs partly project beyond the tip of the tail.

1. Greater hand coverts.
2. Median hand coverts.
3. Alula (thumb).
4. Lesser arm coverts. Several rows.
5. Median coverts. One row.
6. Greater arm coverts. One row.
7. Hand. The outermost part of the wing from the carpal joint to the tip of the wing, consisting of alula, hand coverts and primaries.
8. Arm. The inner part of the wing from the carpal joint to the body, consisting of secondaries and arm coverts.
9. Tertials. The innermost secondaries, which on birds at rest protect and partly cover the flight feathers on the folded wing.
10. Sides of rump.
11. Projection of toes beyond the tip of tail.
12. Secondaries.
13. Primaries (primary flight feathers) most often 10 which are numbered P1 to P10 from the inner one to the outer one.

Underside–feather groups and details

Juvenile Wood Sandpiper. 4.8. NJL.

The underside clearly shows the individual feather groups and details which most waders have in common.

When observing flying birds the underwing often seems plain or simply contrasting dark and pale. However, notice the details which often separate individual species.

In particular, pay attention to details and nuances in feather groups such as axillaries, the markings on the underwing coverts and the undertail coverts. These feather tracts are more or less conspicuously marked but often they are the key to a safe identification.

1. Carpal joint.
2. Median underarm coverts. One row.
3. Greater underarm coverts. One row.
4. Axillaries or armpit-feathers. This feather tract may be uniformly coloured, as for instance on Grey Plover, or barred as on the snipes.
5. Lesser underhand coverts. Several rows.
6. Median underhand coverts.
7. Greater underhand coverts.
8. Shaft streak. On several species the colour of the shaft of the outer primary is different from the rest.
9. Underside of tail.
10. Undertail coverts. In this instance white. Can be coloured or patterned.
11. Vent. The area between belly and tail.
12. Trailing edge of wing.

Head markings

Among waders, the snipes and the Charadrius plovers have the most detailed head markings.

The maze of stripes on the head of the Common Snipe (left), and the Little Ringed Plover (right) illustrate what to pay attention to when trying to identify the bird.

Many of the details and designations are useful when identifying other wader species.

Besides the head markings, a number of additional feather details are shown on the Common Snipe, which are useful for many other waders.

◀ **Common Snipe** (24.9. LG)
1. Upper mandible.
2. Lore. Dark band between base of bill and eye.
3. Supercilium. Narrow, short or long as here.
4. Median crown-stripe. Dark or pale.
5. Lateral crown-stripe.
6. Eye-stripe. The stripe behind the eye.
7. Eye-ring. Ring of tiny pale feathers around the eye.
8. Inner markings on a feather.
9. Tail.
10. Tertials. Note markings or wear.
11. Edge of feather. Plain, notched, barred or spotted.
12. Vermiculated flanks.
13. Fringes. Pale, frayed-looking edgewhich is soon worn off.
14. Breast sides.
15. Malar stripe.
16. Lower mandible.

▶ **Little Ringed Plover** (6.5. LG)
1. Breast-band. Can be broken in the middle.
2. Throat.
3. Base of bill. Inner part of both upper- and lower mandible. Bicoloured in many waders.
4. Nostril and nasal groove.
5. Forehead with black frontal bar.
6. Orbital ring. Coloured ring of skin around the eye.
7. Cheek and ear-coverts.
8. Neck collar.

Feathers and moulting cycle

The feathers serve the purpose of protecting, insulating and camouflaging the bird and of signalling its strength to others during the mating season.

The feathers become worn and are replaced according to a fixed annual pattern, determined by the bird's need for particular groups of feathers at different times of the year. For example, body plumage needs to look at its colourful best at the start of the breeding season.

Juveniles and immature birds

Some of the small-sized species, such as Little Stint and Common Ringed Plover, breed from not quite one year old and have already acquired complete breeding plumage during their first summer. They are distinguishable from adult breeding birds by their retained juvenile flight feathers.

Among some of the larger wader species, which do not breed until they are almost two years old or older, the age of a juvenile can be determined with reasonable ease until the first summer (one year old), when the plumage either resembles the winter plumage or is a pale version of the adult breeding plumage.

During the following autumn, the juvenile moults according to the same schedule as the adult bird and subsequently it is usually impossible to tell it from the non-breeding adult.

Adult

The adult bird changes its plumage twice a year, primarily before and after the migration to the breeding area.

The first moult of the year, which is incomplete, takes place from January to April, during which period the bird moults the worn feathers of head and body together with a varying number of scapulars and wing coverts, while retaining the primaries and secondaries which are to carry it on to the breeding area.

▲ Little Stint, pullus (downy chick).
The chicks of waders wear colours that match their surroundings. The chicks of sandpipers display their so-called 'powder puff' down; the whitish tips to the down feathers.

This apparent sprinkling of powder is visible during the first couple of weeks and provides camouflage against the grass, moss and light-coloured lichens of the tundra. Taymyr, Siberia. Russia, 29.7. JKAM.

▼ Juvenile Temminck's Stint
A textbook example of beautiful and brand new plumage, with pristine, pale-fringed feathers, most clearly seen on the greyish-brown mantle, shoulders, coverts and tertials. 2.8. DP.

▲ Juvenile Dunlin
The first generation of true feathers replaces the down and is quite similar to adult breeding plumage.

However, these juvenile feathers are not particularly durable and are replaced when the juvenile arrives at staging areas such as the Wadden Sea, often half-way to the winter quarters.

At staging areas, fat deposits are re-established straight after the first long journey, and the worn feathers are moulted in favour of a light-coloured plumage, which is less conspicuous on the sand and mudflats.

This still short-billed and short-tailed young Dunlin is about two weeks old, still displaying a downy nape and vent. It will not be fully ready to fly until about 24 days old. 28.7.Norway.AK.

 Several of the larger wader species do not breed until their second or third year and spend their first year in the winter quarters, during which period they do actually moult yet remain in a drab, greyish plumage. Some individuals join the northward trek of the adults, but mainly spend the summer in the roosting areas along the route.

This Bar-tailed Godwit, which is identifiable as a male by its short bill, would not generally breed until its third calendar year as an almost two-year-old bird. Here this bird in its first summer shows hints of copper colour on the sides of the breast and a distinct moulting gap between the new inner four primaries, the upper one still growing, and the outer older ones. Typically, waders shed their worn primaries by starting with the three to five inner ones and then replace the rest in outward progression, so that flying capability is retained. The secondaries are moulted from the outside towards the body. 11.6. The Wadden Sea. LG.

◀ Dunlin, first winter.
At first glance, both juveniles and adults of almost all *Calidris* sandpipers appear to wear a winter plumage of a uniform greyish-brown. On closer inspection, however, and in favourable conditions (particularly in high-quality photos) it is possible to distinguish between different age groups. The bird here is identified as a juvenile on account of the darker unmoulted juvenile scapulars with reddish-brown edges, as well as the dark tips of the greater coverts and of the tertials. 21.1. KBJ.

◀ To the untrained eye, waders look very peculiar duing their two intermediary stages – between winter plumage and breeding plumage on one hand, and breeding plumage and winter plumage on the other. Yet these plumages are very characteristic, and with some degree of practice they become easily recognisable.

This photo shows an adult Curlew Sandpiper, non-breeding to breeding, showing white fringes on the emerging coloured feathers. These are most apparent on the shoulders and are hinted at by a reddish-brown tinge on its neck and breast. The edges will soon wear off, revealing the bright-coloured breeding plumage. 20.5. Greece. DP.

In the *Calidris* sandpipers and other wader families with distinct seasonal plumages, the moulting entails a dramatic change in appearance from a greyish winter plumage to a variegated intermediate stage between summer and winter plumage. This is referred to as adult non-breeding to breeding, with pale fringes which are quickly worn off revealing the full, colourful breeding plumage (adult breeding).

The second moult, which is a complete one, occurs after the breeding season and during the autumn in July to October, when the bird succes- sively sheds and replaces all its feathers, including the worn ones of tail and wings.

In late summer, adult birds in flight show the characteristic moulting gap separating the worn flight feathers from the shorter, growing ones.

At this time the body plumage again changes into a mixture of breeding and non-breeding plumage (adult breeding to non-breeding), until the bird has acquired complete winter plumage (adult non-breeding).

Certain Arctic species, among them the Curlew Sandpiper, have a suspended moult. This means that the body feathers and inner primaries are shed and replaced in the breeding area, after which the moult is postponed until the energy-consu- ming long-distance migration to the tropical winter quarters has been completed.

Uniform plumages

In the wader families that present a more uniform appearance throughout the year, such as curlews and *Tringa* sandpipers, adult breeding and juvenile individuals can be told apart solely by the degree of wear shown in their plumages, when seen on their return journey from the breeding area in late summer and early autumn.

As a general rule the worn adult birds make up the first migration wave, followed by the juveniles with their pristine fresh plumage and more contra- sting feather patterns.

◀ The Wood Sandpiper is among those species with nearly identical plumages all the year round. This individual, photographed on its return migration, is in very worn adult breeding plumage and illustrates the fact that lighter parts of the feathers are less resistant to sunlight and consequently deteriorate and are worn down more rapidly than the darker ones.

Notice also the almost blackish-brown upperside where the white fringes and spots on coverts, scapulars and tertials, so characteristic of fresh plumage, are practically non-existent in late summer. 4.7. LG.

◀ Wood Sandpiper, juvenile.
Compared to the worn adult above, this juvenile neatly shows the difference between the two age groups.

The plumage is fresh with pristine feathers showing the characteristically serrated tertials and pale buff edges and spots that are especially typical of juvenile *Tringa* sandpipers and curlews. 9.8. LG.

Aberrations

A few individuals deviate from the typical moulting schedule, perhaps because they have been injured or infected with parasites. Others may show misshapen bills and legs, because of previous accidents. Other waders show atypical appearance for quite different reasons, as would seem to be the case in the two individuals on this page.

◀ Northern Lapwing, adult breeding. Some individuals have a paler than normal plumage because of leucism. This is a partial or complete absence of eumelanin (black and greyish-brown pigment) and/or phaeomelanin (red-brown pigment). It is usually caused by genetic mutation, but malnutrition can also cause pigment loss. Totally white birds are rare, but partial leucism or dilution less so.

The opposite and much rarer phenomenon is melanism, where the bird has excessive melanin pigment and may be practically jet-black. 22.4. BLC.

◀ Hybrids constitute the rarest variations from normality among waders. By qualified guesswork it is usually possible to identify the species of at least one of the parents on the basis of the most conspicuous features.

This is a presumed Dunlin x Baird's Sandpiper, Pectoral Sandpiper or Sanderling; note the rather long primary projection.

All suggestions are welcome.
Sejerø. Denmark. 31.10. TO.

Feathers and moulting

The moulting cycles of the various wader species form a confusingly varied pattern which has to be understood bit by bit through experience in the field, comparisons of observations, and the study of comprehensive scientific literature. A full knowledge of the stages of moult will give you a far greater insight into the age and life history of each individual bird and so enrich every observation.

Most species moult according to a definite pattern, but within the individual species there are variations related to subspecies, distribution and migration routes.

The photo shows a Common Ringed Plover of the nominate subspecies, *Charadrius hiaticula hiaticula*, which breeds within an area comprising the British Isles, southern Scandinavia, northwest France, the Baltic shores and scattered areas in Poland.

The migration distance from the breeding area to the primary wintering quarters along the European coasts of the Atlantic is a short one. Accordingly the body feathers are moulted in the breeding area or its vicinity, whereas the flight feathers are shed and replaced by new ones continually during late summer and autumn.

For Common Ringed Plovers of the subspecies tundrae, which breeds in northern Scandinavia and along the Siberian and Russian shores, it is essential to save the few accumulated grams of fat from the short summer in the breeding area, to help fuel the energy-demanding southward migration to the winter quarters. So only a few primaries are lost and replaced by new ones before departure.

As for birds with long migration routes, it generally holds true that moulting is postponed completely and the old, well-worn primaries have to carry the bird until it reaches the winter quarters safe and sound, and the plumage can be renewed in a milder climate and with access to plentiful food.

Thus a slightly smaller Common Ringed Plover with retained worn primaries, seen taking a rest on a sandbank on an autumn day, might be a migrant of the tundrae subspecies which has spent its summer among Little Stints, Red Knots, Pomarine Skuas, Gyrfalcons and Arctic Foxes in the Siberian tundra.

The following spread

Dunlin, *Calidris alpina*, is the most commonly observed wader in Europe and has therefore been selected as the species of reference throughout the book.

Once well versed with the size and appearance of the Dunlin, you will be able to compare an unidentified species with this familiar 'template' when you are out wader-watching, and thus build an idea of its relative size and shape and from this perhaps its genus or even species.

The left side of the following spread shows nine plumage types of Dunlins. These variations are shared by most *Calidris* sandpipers, as well as by other wader families with very different seasonal plumages.

Moulting schedule

The right side of the spread presents the plumage terms employed in this book.

Feathers in detail

The line drawings below the moulting schedule illustrate the details of the various feather types, which are characteristic of by far the majority of waders. These types are used for description and identification purposes, especially in the more detailed captions.

For further accounts of feathers, feather groups and related details, see the section on topography (pages 22–27) as well as the captions under illustrations in the species section. There the differences between juvenile and adult and those between worn and fresh feathers, are illustrated and explained.

◀ Common Ringed Plover. *Charadrius hiaticula hiaticula*. First summer. The species breeds at the age of one to two years and this bird, which has migrated to the breeding ground in Denmark, is presumed to be non-breeding, but a summer visitor.

Its age is determined on the strength of the almost worn-down tertials and the heavily worn and faded, yet retained juvenile flight feathers.

On the crown and among the ragged scapulars and coverts, where only the stiff white shafts protrude, a few badly needed fresh greyish-brown feathers are emerging as the first signs of the coming winter plumage. 2.7. LG.

Ad br. With whitish, frayed fringes on the freshly grown scapulars and belly feathers. 17.5. JL.

Ad br. Fresh br plumage with rusty-red upperside and solid black belly patch. 26.5. DP.

Ad br. Very worn br plumage with a few new grey scapulars and coverts. 9.8. LG.

Ad br to non-br. With predominance of new grey winter-feathers on the upperparts. 10.9. LG.

Juv in typical plumage with pale-edged, fresh feathers on the upperside and streaked neck and breast. 3.9. LG.

Post juv. With grey first-winter feathers on mantle and scapulars and juv coverts and tertials. 28.9. LG.

First winter with a few retained cinnamon-edged juv coverts and scapulars. 21.1. KBJ.

Ad non-br. Has plain brownish-grey plumage with characteristic large and uniform coverts. 20.1. KBJ.

Presumed first summer with grey winter feathers and just a hint of rusty-red on ear-coverts and lores. 9.5. HS.

Plumages	Description	Period of plumage	Calendar year/Age
Pullus	Downy young	Until first juvenile feathers emerge	1 Cy
Juvenile	Young bird in its first plumage	June to September	1 Cy
Post juvenile	Young bird moulting to first winter	August to November	1 Cy
First winter	First winter plumage	October to April	1 Cy-2 Cy
Moult to first summer	Immature moulting to first summer	February to April	2 Cy
First summer	First summer plumage	February to September	2 Cy/One year old
Moult to second winter	Moulting to second winter plumage	June to October	2 Cy
Second winter	Second winter plumage	August to April	2 Cy-3 Cy
Moult to adult breeding	Moulting to adult breeding plumage	February to May	3 Cy
Adult breeding	Adult breeding plumage	April to September	3 Cy/Two years old
Adult breeding to non-breeding	Adult moulting to winter plumage	July to November	3 Cy
Adult non-breeding	Adult in winter plumage	October to April	3 Cy-4 Cy

Shaft streak Often dark, but can be distinct and pale, particularly on the outer primaries.

White-tipped Most distinct on fresh feathers where the tips are not yet worn.

Chevron Mostly seen on breast- and flank feathers. Called vermiculated if the marking is thinner and less distinct.

Subterminal bar Often seen on scapulars and arm-coverts.

Subterminal anchor Often seen on scapulars and arm-coverts.

Edged Pale edge on, for example, the outer side of scapulars in some snipes.

Spotted edge Pale spots along the edge of a darker feather, as on scapulars of Spotted Redshank.

Striped edges Separated, short dark stripes on paler feathers, as on scapulars of Common Redshank.

Barred Often on tertials, for example on *Tringa* sandpipers and curlews as well as on axillaries and tail feathers.

Serrated or notched Typical on *Tringa* sandpipers, where the pale notches are faded and gradually worn off.

Internally marked Asymmetrical markings which do not extend to the edge of the feather. Often seen on snipes.

Dark-centred Common feather pattern on many waders, particularly on the scapulars.

Silhouettes and jizz

The size, shape and habitual movements of a particular bird create a general impression or 'jizz', which helps separate it from other species.

By noting its silhouette and behaviour for a few seconds we obtain useful information to help identify it, as well as assess its 'mood' and perhaps predict its behaviour.

The silhouettes shown in this book should be seen as valuable tools, providing a shortcut towards reliably telling apart the more similar wader species.

With a little practice at studying silhouettes, you can become skilled at quickly assigning birds to the correct family and often to species, even without the use of binoculars.

On the following two spreads the silhouettes of the most characteristic European waders are presented.

The silhouettes to the left represent waders likely to be present in the Wadden Sea on a spring morning.
See whether you can determine their species, on the basis of your own experience, and by referring to the silhouettes depicted in this book.

The answers are below from left to right:

Red-necked Phalarope, *Phalaropus lobatus*.
Black-tailed Godwit, *Limosa limosa*.
European Golden Plover, *Pluvialis apricaria*.
Common Redshank, *Tringa totanus*
Peregrine Falcon, *Falco peregrinus*.
Dunlin, *Calidris alpina*, flock, small group in the background, and roosting party.
Pied Avocet, *Recurvirostra avocetta*.
European Golden Plover, *Pluvialis apricaria*, foraging.
Red Knot, *Calidris canutus*, flock.
Northern Lapwing, *Vanellus vanellus*.
Broad-billed Sandpiper, *Calidris falcinellus*.
Common Greenshank, *Tringa nebularia*, foraging.
Eurasian Oystercatcher, *Haematopus ostralegus*.
Ruff, male and female, *Calidris pugnax*.
Eurasian Curlew, *Numenius arquata*.
Bar-tailed Godwit, *Limosa lapponica*, flock.

Eurasian Curlew
Numenius arquata ▲▼

Whimbrel
Numenius phaeopus ▲▼

Black-tailed Godwit
Limosa limosa ▲▼

Bar-tailed Godwit
Limosa lapponica ▲▼

Grey Plover
Pluvialis squatarola ▲▼

Spotted Redshank
Tringa erythropus ▲▼

Common Greenshank
Tringa nebularia ▲▼

Common Redshank
Tringa totanus ▲▼

Eurasian Oystercatcher
Haematopus ostralegus ▲▼

Black-winged Stilt
Himantopus himantopus ▲▼

Pied Avocet
Recurvirostra avocetta ▲▼

Eurasian Stone-curlew
Burhinus oedicnemus ▲▼

Eurasian Woodcock
Scolopax rusticola ▲▼

Great Snipe
Gallinago media ▲▼

Common Snipe
Gallinago gallinago ▲▼

Collared Pratincole
Glareola pratincola ▲▼

Red Knot
Calidris canutus ▲▼

Ruff
Calidris pugnax ▲▼

Ruddy Turnstone
Arenaria interpres ▲▼

Eurasian Dotterel
Charadrius morinellus ▲▼

Terek Sandpiper
Xenus cinereus ▲▼

Dunlin
Calidris alpina ▲▼

Curlew Sandpiper
Calidris ferruginea ▲▼

Sanderling
Calidris alba ▲▼

Common Ringed Plover
Charadrius hiaticula ▲▼

Marsh Sandpiper
Tringa stagnatilis ▲▼

Green Sandpiper
Tringa ochropus ▲▼

Common Sandpiper
Actitis hypoleucos ▲▼

Broad-billed Sandpiper
Calidris falcinellus ▲▼

Little Stint
Calidris minuta ▲▼

Jack Snipe
Lymnocryptes minimus ▲▼

Red-necked Phalarope
Phalaropus lobatus ▲▼

Plates for comparison

Many small and medium sized waders closely resemble each other, making identification difficult.

During the summer season most species display an upperside variegated in black, white and brown, with reddish and golden tinges and in the winter season many species have nearly uniform greyish-brown plumage.

Thus the similarity between the various species of *Calidris* sandpipers may seem overwhelmingly confusing and for the unpractised eye a mixed flock of waders on their autumn migration may well present a daunting and frustrating task when it comes to identification of the various species.

The plates on the following pages show the waders among which the similarity between the species is greatest. They also indicate the most important field and flight features by means of arrows and supplementary captions.

These plates are intended to act as 'revision' material prior to a field trip and also as a quick reference in cases of doubt.

The comparison plates are followed by five illustrations showing bird flocks in the field.

These pictures show the birds in their natural surroundings as they are seen through binoculars, and offer a more realistic impression of waders in their habitat, as well as allowing for the comparison of sizes, which is an important factor during species determination in the field.

Four species of waders gathered around a small tussock during a rising tide in the Wadden Sea.
From left to right:
Red Knot, *Calidris canutus*
Grey Plover, *Pluvialis squatarola*
Dunlin, *Calidris alpina*
Common Redshank, *Tringa totanus*
All birds are in worn adult breeding plumage. 9.8. LG.

Common Ringed Plover. Ad ♂. 7.7. LG. Size of Dunlin. Recognised from the other European plovers by the orange-yellow bill with black tip (1), the face mask with broad, black band over fore-crown and broad, black line from bill to hind crown (2), broad, black neck ring (3), and orange-yellow legs (4).

Semipalmated Plover. Ad ♂. 6.6. DP. Rare vagrant from North America. Told from Common Ringed Plover by a combination of distinct yellow orbital ring (1), faint whitish spot over the back of the eye (2), narrow white band on forehead (3), narrower neck ring (4) and short, petite bill (5). Additionally, the two safest field marks are call and the partially webbed toes.

Killdeer. Ad. 6.6. DP. Rare vagrant from North America. Has distinct, red orbital ring all year round (1), double, black neck and breast-bands (2), long tail (3), cinnamon-coloured uppertail coverts (4), and long, flesh-coloured legs (5).

Little Ringed Plover. 6.6. L.G. Smaller than Common Ringed Plover and easily recognised from it and from Kentish Plover by the bright yellow orbital ring (1) and the flesh-coloured legs (2). Breeds primarily at inland habitats such as gravel pits and lakes.

Kentish Plover. Ad ♂. 26.5. LG. The palest of the European plovers. Recognised by the more or less ochre crown (1), the weak, all-black bill (2), the black, narrow and broken breast band (3), and the greyish-black legs (4). The adult ♀ in br plumage is less distinctly marked and paler than the ♂.

Eurasian Dotterel. Ad ♀. 18.6. DP. Larger than Common Ringed Plover. In ad br plumage unmistakable with orange breast grading to blackish-brown on the belly. The prominent white supercilium (1) and the narrow breast band (2), also seen in juv plumage, separates the species from juv *Pluvialis* plovers. Roost and breeds at inland localities.

Lesser Sand Plover. Ad ♂. 2.5. DP. Rare vagrant from Asia. Differs foremost from the similar Greater Sand Plover by the bill, which has a faint, saddle like hollow near the middle (1) continuing into an elongated, bulging shape (2) towards the pinched tip. Furthermore, the legs are situated behind the central point of the body (3). Five subspecies and considerable plumage variation.

Greater Sand Plover. Ad ♂. 6.6. EFH. Slightly larger and more powerful than Lesser Sand Plover, with flatter head and stronger, dagger-like bill (1). The bird seems more symmetrical with the eye placed in the middle of the head and the legs placed exactly below the central point of the body. The shape may recall a small Grey Plover. Three subspecies.

Caspian Plover. Ad ♂. 10.4. KBJ. Elegant and long-billed with upright, golden plover-like stance and large, dark eyes in a pale head. The reddish breast band, greyish-brown on the ♀, is bordered by a narrow, black band towards the belly (1). Because of the long legs the bird seems large, but its body is not larger than Common Ringed Plover's. Rare breeder in south-eastern Europe.

Common Ringed Plover. Juv, still partially downy. 22.7. NLJ. Appears neatly 'scaled' with narrow, pale edges on upperside (1), pale-edged tertials (2), and an orange spot on the base of the lower mandible (3). Does not have yellow orbital ring like juv Little Ringed Plover. Does not have black legs and broken breast band like juv Kentish Plover.

Semipalmated Plover. Juv. 15.10. HS. Very similar to juv Common Ringed Plover, but unlike that species it has the suggestion of a yellow orbital ring (1), also less white over the far part of the eye (2), a shorter and heavier bill (3), and partially webbed toes (4). Also try to note the call, which differs from that of Common Ringed Plover.

Killdeer. Post juv. 31.10. DP. Rare vagrant from North America. Easily recognised, also as juv, by the red orbital ring (1), double black neck and breast-bands (2), long tail (3), and cinnamon-coloured uppertail coverts, partly hidden in this photo (4).

Little Ringed Plover. Juv. 5.7. JL. Easily told from the other European plovers by the bright yellow orbital ring (1), and the brown subterminal bands below the narrow, pale fringes on the upperside feathers (2). Also note an orange spot at the base of the lower mandible (3), as in Common Ringed Plover.

Kentish Plover. Juv. 30.8. SP. The palest and most greyish-white of the three common ringed plovers. Has grey-brown upperside with white fringes. Differs from juv Little and Common Ringed Plovers by the all-black bill (1), the blackish-grey legs (2) and the white neck with no black band (3).

Eurasian Dotterel. Juv 19.8. JL. Somewhat larger than Common Ringed Plover, more stout and with deep breast. Could be confused with juv *Pluvialis* species because of coarsely mottled upperside (1), but always distinguishable from these by the prominent, white-beige supercilium (2), the white band across the breast (3), the beige flanks (4) and the brown-beige edges of the tertials (5).

Lesser Sand Plover. Juv. 22.10. AJ. Differs from the quite similar, slightly larger Greater Sand Plover by the bill, which has a slight saddle-like hollow near the middle (1) continuing in an elongated, bulging shape (2) towards the pinched tip. Furthermore the legs are situated behind the central point of the body. Five subspecies and considerable plumage variation.

Greater Sand Plover. Juv. 27.8. DP. Differs from the slightly smaller and quite similar Lesser Sand Plover in its more powerful appearance, due to flatter shape of head and stronger, dagger-like bill (1). The bird seems more symmetrical with the eye placed in the middle of the head and the legs placed exactly below the central point of the body. Three subspecies.

Caspian Plover. Juv.12.9. HJE. Similar size to Common Ringed Plover, but more elegant, slimmer and with longer legs. Differs from the other juv *Charadrius* plovers and juv Dotterel by the uniform, brown upperside with broad, beige feather edges (1), and the long legs (2).

Grey Plover. Ad non-br to br. 16.5. HS. Medium-sized wader, about the size of Northern Lapwing. A little larger and with heavier bill (1) than the other *Pluvialis* plovers. In spring and autumn the contrastingly black and white upperside is diagnostic (2). This individual is moulting into br plumage.

Grey Plover. Ad ♂ br. 17.6. DP. The largest *Pluvialis* plover, with stout body and heavy bill (1). Unmistakable in the black/white br plumage with pure white vent (1) contrasting with back belly.

Grey Plover. Juv. 21.9. JL. Told from the other *Pluvialis* species by size, heavy bill (1) and the pure white vent (2). Note that juv in fresh plumage may have a golden wash to the upperside, like the other juv *Pluvialis* species.

European Golden Plover. Ad non-br to br. 5.4. EFH. The most common *Pluvialis* plover in Europe. Somewhat smaller and with a milder expression than Grey Plover. Has more or less golden upperside all the year (1) and in the summer season black markings on head, breast and belly (2). ♀ duller than ♂.

European Golden Plover. Ad ♂ br. 12.5. JL. Elegant and distinctive in br plumage with black markings on head, neck and belly. The long white demarcation line from forehead to vent (1) separates the species from the longer-legged American Golden Plover. The similar Pacific Golden Plover is noticeably smaller with longer legs, particularly the tibia.

European Golden Plover. Juv. 5.8. JL. Standing birds are told from juv Grey Plover by having smaller body, weaker bill (1) and by the spotted underside, which is white-beige with narrow brown streaks in Grey Plover. Differs from juv American and Pacific Golden Plovers by differences in markings on upperside, length of legs, primary projection and length of tertials.

American Golden Plover. Ad br to br. 13.8. TH. Rare vagrant from North America. Distinguishable all year round from European Golden Plover on brownish, not white underwing (1) and on longer legs. Outside the br season differs from Pacific Golden Plover in differences in primary projection and length of tertials. See description of species.

American Golden Plover. Ad br. 15.6. KK. Rare vagrant from North America. Unmistakable with all-black underside and broad white border (1) from forehead to front of wing at the shoulder. On both European and Pacific Golden Plovers the white border continues to the lower flanks just behind the legs. Considerably longer-legged (2) than European Golden Plover.

American Golden Plover. Juv. 1.12. LK. Rare vagrant from North America. Separated in flight from European Golden Plover on grey-brown underwing (1), which is pale greyish-white on Golden Plover. Very similar to the marginally smaller and slightly more long-legged juv Pacific Golden Plover, which also has brownish underwing, but larger spots on golden upperside.

Grey Plover. Ad ♂ br. 9.5. BLC. Unmistakable in its black/white br plumage. Notice the black axillaries (armpit feathers) (1), which are diagnostic and are present in all plumages.

Grey Plover. Juv. 21.9. JL. The diagnostic black axillaries (1) stand out against the white underwing and provide an infallible identification mark in all plumages. Furthermore, juv are distinctly streaked with a yellowish-beige tinge on head, breast and belly.

Grey Plover. Ad non-br. 7.1. DP. Distinguished from the other *Pluvialis* species in non-br plumage by its colder, grey-brown look without golden feathers and by its more bulky body and head as well as heavier bill (1).

European Golden Plover. Ad ♂, br to non-br. 1.8. JL. In br plumage differs from American and Pacific Golden Plovers by greyish-white underwing (1) and specifically from American Golden Plover by white border (2) extending from forehead to vent as well as on white undertail coverts (3). On American Golden Plover the white border extends only to edge of wing, in line with shoulder.

European Golden Plover. ♂ br. 24.4. JL. Has the most golden upperside (1) of the three similar golden plover species. The golden upperside and the distinct pale wing-bar (2) combined with the long white border from forehead to vent and the pale, almost white underwings make this species easy to identify in br plumage.

European Golden Plover. Ad non-br. 1.12. KBJ. Shorter-legged and far more golden on the upperside than its two close relatives, American Golden Plover and Pacific Golden Plover, which do not normally occur in Europe in winter. Told from juv by pure white belly (1), and ad tertials (2), which are not divided by a narrow black line at the tip. Almost impossible to tell from first winter after mid winter.

Pacific Golden Plover. Ad br. 5.5. HJE. Rare vagrant from Asia. Most easily told from the similar European Golden Plover by appreciably smaller body and longer legs (1). On standing birds the vital details are the primary projection and the tertials (2). See description of species.

Pacific Golden Plover. ♂ br. 28.4.HS. Rare vagrant from Asia. Told from European Golden Plover and American Golden Plover respectively by the whitish-brown underwing with pale brown axillaries (1), and the long white border from forehead to vent (2).

Pacific Golden Plover. Juv. 10.9. MV. Rare vagrant from Asia. Standing individuals differ from European Golden Plover and American Golden Plover on smaller body, longer legs (1), golden upperparts (2) and conclusive details in primary projection and length of tertials. See description of species.

Dunlin. Ad br. 24.7. OJL. White wing-bar, narrow on the arm and broad on the hand (1). Uneven transition between primaries and secondaries, a so called moulting gap (2) shows that the bird has recently shed some secondaries, and the new ones have not yet attained their full length. Also note rusty-red scapulars (3), black, medium-long and slightly curved bill (4), black belly patch and densely streaked breast (5).

Dunlin. Juv. 21.8. JL. Almost all *Calidris* species have white sides to the rump, and a grey tail with a dark central stripe (1). Juv do not show moulting gaps along the trailing edge of wings (2). Compared to ad shows only few feathers with rusty red fringes (3). Has paler head and throat area than ad (4). Almost all *Calidris* species show a white wing-bar of variable width and length (5).

Dunlin. Juv. 16.8. JL. On flying Dunlins and most other *Calidris* species, the toes more or less reach the tip of the tail (1). Almost all *Calidris* species have pure white underwing coverts (2). Unlike ad, has varying degree of ochrous hue on head and upper breast (3) Streaking on breast merges with black spots on the flanks (4); on some individuals these black spots even suggest a black belly patch (5).

Curlew Sandpiper. Ad br. 24.7. JL. Toes protrude beyond the tip of the tail (1). In br plumage shows diagnostic white rump, spotted with blackish-brown (2). Has slim, quite long bill curving towards the tip (3). Brick-red head and underside, of similar colour to Red Knot and Bar-tailed Godwit.

Curlew Sandpiper. Juv. 25.8. JL. non-br and juv have pure white rump without the dark stripe characteristic of other *Calidris* species (1). The only other *Calidris* with a pure white rump is the rare American vagrant White-rumped Sandpiper. Has uniform greyish brown upperparts with broad, pale fringes (2).

Curlew Sandpiper. Juv. 16.8. JL. Unlike Dunlin, very pale with white rump and belly (1), and a diffuse ochre-beige tinge on neck and upper breast. Faintly streaked on sides of breast (2).

Sanderling. Ad br. 25.5. JL. More compact than Dunlin. White belly in all plumages (1). Broad, white wing-bar (2). Short, black and straight bill (3). Mottled reddish, black and white on throat and neck (4).

Sanderling. Juv. 19.8. EFH. Has grey tail feathers like the majority of *Calidris* species (2). Has greyish upperparts with characteristic chequered patterns on scapulars and mantle (2). *Calidris* species typically have the white rump intersected by a dark, central stripe, consisting of darker uppertail coverts that in most cases blend with the grey tail (3).

Sanderling. Non-br. 21.3.BLC. In all plumages Dunlin-sized, robust and short-billed. Differs from juv in having uniform, pale grey back (1). Has diagnostic, contrasting black and white wing markings with very dark leading edge of inner wing (2).

Little Stint. Juv. 25.8. SSL. Little Stint and Temminck's Stint are smaller than Dunlin, and the only two of the seven so-called stints that occur commonly in Europe. The other five are American or Asian vagrants. The upperparts are marked with black, white and rusty brown. One of the most conspicuous juv plumages. Has white, almost parallel streaks on back (1). Short, black almost straight bill (2).

Temminck's Stint. Br. Song-flight. 11.6. HS. One of the seven stints, and a similar size to Little Stint. Unusual in having grey and drab plumage all year round. In the summer season it is thus easy to identify from Little Stint and the other *Calidris* species. The only *Calidris* species with pure white outer tail feathers (1). Has greenish-yellow legs; Little Stint has greyish-black legs (2).

Broad-billed Sandpiper. Ad br. 27.5. HS. Quite dark in the summer season. The upperside may even appear blackish with white lines when the bird is worn in late summer. The toes reach beyond the tip of the tail, contrary to most *Calidris* species (1). Has diagnostic head-markings with white supercilium and pale crown stripes (2). The bill has a diagnostic kink near the tip (3).

Red Knot. Ad br. 2.6. JL. Large *Calidris* sandpiper, considerably larger than Dunlin. Finely barred rump (1), short, quite heavy and straight bill (2), brick-red head and underside, like ♂ Bar-tailed Godwit but is smaller with shorter legs and bill (3).

Red Knot. Juv. 17.8. JL. Large *Calidris* species. Has finely barred rump (1), and greyish-brown upperside with broad scaly fringes (2). Initially has reddish-ochre wash on head, breast and belly, but this soon fades (3).

Purple Sandpiper. 13.3. HS. Has characteristic, blackish-grey upperside all year round. Tail with broad, dark central bar contrasts with black-streaked white sides of rump (1). Greyish-black head (2). The bill is dark yellow at base and darker towards the tip (3). Coarsely spotted breast and streaked flanks, belly and vent (4).

Ruff. Non-br. 29.3. EFH. Presumed ♀ (♀ a quarter smaller than ♂) because the shorter body means the feet reach further beyond the tail. Toes or feet always visible behind tail (1). Often shows characteristic white, U-shaped pattern on rump, if the central bar does not extend to the tail feathers (which is sometimes the case) (2). Dark, short bill often with a pale area around the base of bill (3).

Ruff. Juv. 5.8. JL. Has characteristic U-shaped white marking on rump, or V-shaped as here, where the central bar extends to the tail feathers (1). In all plumages shows a thin, pale wing-bar on the secondary coverts (2). Upperside strongly scaled with blackish-brown feather centres (3). Buff to ochrous-beige wash on head and underside (4).

Ruddy Turnstone. Juv. 4.9. HS. Has diagnostically patterned upperside in all plumages. Broad white wing-bar (1). White tail with broad subterminal band (2). White wedge on wing edge, and along the scapulars, also white stripe on back (3). Dark, short, thick-based bill (4). White shaft-streaks on primaries (5).

Dunlin. Br to non-br. 9.8.LG. Tips of the folded wings just reach the tip of the tail (1). Rusty-red and black shoulder feathers in contrast to new, grey mantle feathers and greater coverts (2). Rather long bill of variable length and with gently curved tip (3). The only *Calidris* species with all-black belly patch (4). Black legs (5).

Curlew Sandpiper. 11.7. DP. Long-winged with primaries projecting past the tip of tail (1). Bill longer and more curved than Dunlin's (2). Pale area around base of bill (3). Dark brick-red face, neck, breast and belly (4), longer-legged than Dunlin (5).

Broad-billed Sandpiper. 29.5. DP. Fairly short tail, with tips of primaries reaching tip of tail (1). Dark upperside with large, blackish-brown feather-centres (2). Bull-necked (3). Characteristic head pattern with dark lores, long, pale supercilium and white crown stripes (5), diagnostic kink at tip of bill (5), triangular dark marks on breast (6).

Sanderling. 31.5. DP. Upperside spangled in orange, rusty-red and black. Some individuals can be very pale, almost black/white (1). Short, straight bill (2), and orange-red throat, neck and breast with fine black spots and streaks (3). The only *Calidris* species without a hind toe (4).

Little Stint. 20.5. DP. One of the seven stints which is significantly smaller than Dunlin. Broad rusty-red edges on tertials, shoulder feathers and coverts (1). Short, black almost straight bill (2), white throat (3), and dark, blackish-grey legs (4). The rusty-red feather edges and the white throat distinguish it from Red-necked Stint, a rare Asian vagrant.

Temminck's Stint. 11.6. HS. Significantly smaller than Dunlin. All year round the dullest and most grey *Calidris* species, but in br plumage has scattered cinnamon-brown, black-centred feathers (1). Short, olive-tinged brownish-black bill with characteristic upturned lower edge of lower mandible (2), and greenish-yellow legs (3). The only *Calidris* with pure white outer tail feathers (4).

Purple Sandpiper. 1.6. DP. Dark grey and stocky with deep breast. The blackish-grey plumage is adorned with rusty-red feather edges on mantle and scapulars in the br season (1). Bicoloured, dark yellow bill fading into black (2), spotted breast and streaked flanks and belly (3), and dark yellow legs (4).

Red Knot. Islandica. 6.6. DP. Robust and stockily built *Calidris*, a great deal larger than Dunlin and Curlew Sandpiper. Variegated upperside in orange-red with anchor-shaped markings on the scapulars (1), short, straight black bill (2), orange-red face and underside (3), and greyish-green legs (4).

Ruff. Ad ♀. 5.5. LG. ♀ are three-quarters the size of ♂ and with rounder heads; ♂ have more triangular head shape. Long, often flutteringly 'loose' tertials and scapulars (1). Scaly-looking upperparts, with black feather centres and rust-coloured barring (2). White around base of bill (3). Normally orange-yellow legs and feet, more rarely brownish or greyish-green (4).

Dunlin. 9.8. LG. Note the all fresh plumage, with no worn feathers and with broad, pale fringes which is typical for all juv *Calidris* species. Looks like a toned-down ad with narrow, white V-marking on back and rusty-red tinge on head, breast and coverts (1). Medium-long bill with slight bend at tip (2), streaked breast (3), and variable belly patch which can look dingy and may consist of spots or streaks (4).

Curlew Sandpiper. 7.9. HS. Easy to tell apart from other juv *Calidris* species by scaly look to back, due to pale fringes (1), medium-long, slightly curved bill (2), ochre-beige, diffuse wash on head and upper breast (3), dark greyish-green legs (4), and pure white underside without any trace of black (5).

Broad-billed Sandpiper. 20.8. HS. At a distance easily mistaken for the slightly larger Dunlin. Dark upperside with white 'V' on back as with Little Stint (1). Characteristic head pattern with dark lores, long pale supercilium and white crown stripes (2), deep bill with diagnostic kink at tip (3), white belly and flanks (4).

Sanderling. 19.8. BLC. A very pale juv *Calidris*. Recognised by chequered shoulders and heavily streaked mantle (1), dingy, beige breast band (3), black legs and feet, lacking hind toe (4), and characteristic black carpal joint, though this may be hidden by the breast feathers (5).

Little Stint. 3.9. LG. Markedly smaller than Dunlin, and with conspicuously well-marked plumage. Tertials and greater coverts with broad cinnamon coloured edges (1), well defined white 'V' on back (2), short, straight and black bill (3), rusty-beige spots on sides of breast (4), and dark, greyish-black legs (5). Often seen in smaller parties, usually together with Dunlins, when size difference is tangible.

Temminck's Stint. 16.8. JL. One of the seven stints which are markedly smaller than Dunlin. Distinct buff-beige fringes on the grey-brown upperside (1). Short, dark, olive-tinged brownish-black bill with characteristic upturned lower edge of lower mandible (2). Washed-out beige-brown breast (3). Greenish-yellow legs (Little Stint is black-legged) (4). Often solitary or a few together.

Purple Sandpiper. 10.8. SD. Fluffed up to keep warm. Has characteristic, blackish-grey upperside with broad white 'V' on back and bronze-coloured fringes on upper shoulder feathers (1). Bicoloured bill, dark yellow at base and blackish-brown towards tip (2). Greyish-white, distinctly streaked breast and belly (3). Greenish-yellow legs (4). Full juv plumage is only seen near br grounds.

Red Knot. 22.8 BLC. Robust and stockily built *Calidris*, a great deal larger than Dunlin. Besides size easily told from other *Calidris* species by greyish upperparts with white fringes and narrow, dark terminal bands (1), short, straight black bill (2), reddish-ochre wash on head, breast and belly, not always as prominent as here (3), and greenish-yellow legs (4).

Ruff. 24.9. LG. ♂. Sexed by large body and triangular head shape. ♂ are approximately 25% larger than ♀. Long, often flutteringly 'loose' tertials and scapulars (1). Large, pale fringes on the greyish-brown upperparts (2). Long-necked, particularly noticeable in 'giraffe-posture' when alert (3). Dark, slightly curved bill (4), beige wash on neck (5), and greyish-green legs (6).

Dunlin. Ad non-br. 20.01. KBJ. Medium-sized *Calidris* with medium-long legs, black bill and legs. The commonest *Calidris* species in Europe all year round. Like almost all its relatives it is greyish-brown above and whitish below. Medium-long, faintly curved bill (1). Dark breast sides with weak streaking (2).

Dunlin. First winter. 21.01. KBJ. Very similar to ad non-br but differs in having retained juv coverts (1), and retained upper scapulars with blackish centres and rusty-brown fringes (2).

Sanderling. First winter. 29.12. KBJ. Very pale *Calidris* with bright, white breast and belly. Seen sprinting along the surf on sandy beaches. Only *Calidris* with no hind toe (1). Pure white underside (2), black carpal patch, here covered by breast feathers (3), and short, strong and straight bill (4). Age determined by retained juv pale edged coverts (5).

Sanderling. Ad non-br. 7.1. DP. Strikingly pale *Calidris* with snow-white breast and belly. Seen in smaller parties running along the surf on sandy beaches. In flight shows very dark front of wing, which on standing birds are noted as a characteristic, dark carpal patch contrasting to the very pale body (1).

Red Knot. First winter. 24.1. HS. Stocky *Calidris* with relatively short legs. Clearly larger than Dunlin. Greyish-green legs (1). Aged on retained, juv coverts and lower scapulars (2). Short, straight and black bill (3).

Red Knot. First summer. 14.4. LG. Breeds when two or three years old and primarily spends the summer season in the winter quarters. Very similar to ad non-br. Aged on retained, worn and faded juv median coverts with remnants of dark subterminal bands (1).

Purple Sandpiper. Ad non-br. 27.1.LG. Medium sized and robust *Calidris* with characteristic, blackish-grey plumage all year round. Greenish-yellow to dark yellow legs and feet (1). Bicoloured bill, dark yellow at base shading into blackish-brown towards the tip (2). Age determined on the large scapulars and coverts with ill-defined, but quite broad pale fringes (3).

Purple Sandpiper. First winter. 7.1. LG. Aged on juv tertials and coverts with narrow and well defined white edges and fringes (1).

Ruff. Ad ♀. 27.1. LG. Largest *Calidris* species, size of Common Redshank. ♀ a quarter smaller than ♂. Large, 'loose' feathers (1). Often has white patch around base of bill (2). Bicoloured bill, reddish at base, dark at tip (3). Reddish-yellow to greyish legs (4). Sexed on size and rounded shape of head.

Temminck's Stint. 17.3. HS. This and Little Stint are the two only stints commonly seen in Europe, though both are scarce in the winter season. In non-br plumage very similar to Little Stint. Greenish to yellowish-green legs, which often look dark (1). Contrary to Little Stint it has characteristic upturned lower edge of lower mandible, though here partly obscured by mud (2).

Little Stint. 3.3. KBJ. With Temminck's Stint, the only stint commonly seen in Europe, albeit scarce in the winter season. In non-br plumage very similar to Temminck's Stint. Unlike the latter it has black legs (1), and straight bill without upturned lower mandible.

Curlew Sandpiper. 12.11. ISA. First winter. Most similar to Dunlin, but differs by the longer and more curved bill (1), and more attenuated body due to longer wings. Aged on juv coverts, scapulars and tertials (2). Rare in Europe in the winter season.

Spotted Redshank. 5.3. HS. Large, long-legged and slender *Tringa* species with long bill. May be mistaken for Common Redshank. Has medium-grey crown and pronounced white supercilium extending to base of upper mandible (1). Long, slim bill with slight hook at tip and red base of inner part of lower mandible (2), white unstreaked belly (3), and long red legs (4).

Common Redshank. *T. t. totanus.* 10.1. HJE. The nominate form, some of which winter in southern Europe. Medium-sized *Tringa* and the commonest *Tringa* species in Europe. Closest to Spotted Redshank in appearance. Medium-long, bicoloured bill, orange-red on inner half, black on outer (1). Faintly streaked underparts (2). Long orange red legs (3).

Common Redshank. *T. t. robusta.* 'Icelandic Redshank'. 17.1. LG. Primarily breeds in Iceland and winters along the north-western coasts of Europe. Differs from the nominate form in having very dark upperparts (1) and more or less heavily streaked underside (2).

Ruddy Turnstone. First winter. 7.11. HS. Short-legged and stocky, medium-sized wader, somewhat larger than Dunlin. Short, greyish-black wedge-shaped bill (1). Greyish-black breast (2). Orange legs all year round (3). Aged on retained juv coverts and scapulars (4).

Common Greenshank. Ad non-br. 19.2. HHL. Large, powerful *Tringa* with long legs. Bill greyish and slightly upcurved with dark tip (1). Greyish to greyish-green legs (2). Aged by the ad tertials (3). Contrary to Marsh Sandpiper it has clearly streaked crown, nape and cheeks (4).

Marsh Sandpiper. Ad non-br. 5.2. HJE. Quite similar to Common Greenshank, but is smaller, more slender and elegant. Contrary to Common Greenshank it has evenly thin and straight bill (1). Green to yellowish-green legs (2) and very pale head with almost unstreaked pale grey crown (3).

Dunlin. 17.5. JL. Medium-sized *Calidris* and the most common in Europe all year round. The only *Calidris* species with black belly patch in the summer season (1). Black bill with slightly down-curved tip suggesting the bill of Curlew Sandpiper (2). In spring the black belly patch has white fringes which are soon worn off.

Broad-billed Sandpiper. 28.4. HS. Medium-sized *Calidris*, about the same size as Dunlin, but more round-shouldered and often with 'bull necked' appearance. Shows characteristic head markings with split supercilium (1). Bill similar to that of Dunlin but deeper and more thick-based with diagnostic 'drop-like' kink at tip (2).

Little Stint. 28.4. HS. One of the two European stints which are noticeably smaller than Dunlin. Unmistakable with its orange-red upperside which becomes even more red when the pale fringes wear off during spring. Differs from the rare vagrant Red-necked Stint by the white throat (1) and the cinnamon-coloured fringes on tertials and coverts (2).

Purple Sandpiper. 5.4. HS. Medium-sized and stout *Calidris* with characteristic blackish-grey plumage all year round. Dark yellow to greenish-yellow legs and toes (1). Bicoloured bill, dark yellow at base and blackish-brown at tip (2). In the winter season typically encountered on rocky beaches, harbour piers and breakwaters.

Sanderling. 1.5. NLJ. In winter season the Sanderling is the lightest *Calidris* species. In summer season, only the snowy-white belly is retained (1) and the rest of the plumage is reddish with more black-and-white admixed. The only *Calidris* with no hind toe. Seen in small parties sprinting along the tide-line on sandy beaches. The grey winter-feathers are moulted and replaced by reddish ones (2). Short straight bill (3).

Temminck's Stint. 24.4. HS. One of the two European stints which are noticeably smaller than Dunlin. In the summer season with black-centred reddish-beige scapulars (1) in the otherwise grey plumage, the hallmark of the species. In the winter season differ from the similar Little Stint by the upturned tip of the lower mandible (2) and the greenish-yellow legs (3).

Curlew Sandpiper. 20.5. DP. Similar in size to Dunlin, but more looks more attenuated due to its long wings (1). Recognized from all other *Calidris* species by the relatively long and slightly curved bill (2). Colour-wise similar to Red Knot in br plumage, but considerably smaller.

Red Knot. 6.4. KK. A short-legged, compact and stoutly built *Calidris* considerably larger than Dunlin. Acquires deep brick-red breeding plumage like that of Bar-tailed Godwit and Curlew Sandpiper. Told from these on size, bill length and shape (1) and by the short greyish-green legs (2).

Ruff. ♀. 15.4. JL. Told from the other *Calidris* species and waders on a combination of behaviour, typically large and "loose" scapulars and tertials (1) as well as the leg-colour which in the summer season is orange-yellow (2). The ♂ is normally one fourth larger than the ♀ and has a spectacular br plumage.

Dunlin. 10.9. LG. Medium-sized *Calidris* sandpiper with medium-long bill and legs. The most common *Calidris* in Europe all year round and (in br plumage) the only one with a black breast-and belly patch. Under the emerging grey winter feathers (1), the black and orange remnants of the br plumage can be seen. The black belly patch is starting to disappear.

Broad-billed Sandpiper. 4.9. HS. Medium-sized *Calidris* about size of Dunlin, but more round-shouldered with 'bull necked' appearance. Shows characteristic head markings with split supercilium (1) and pale crown-stripe (2), and bill has diagnostic 'drop-like' kink at tip (3). The transitional post-br plumage is more contrasting due to new pale grey winter feathers (4) and a few worn and faded flight feathers (5).

Little Stint. 7.11. KBJ. In the transitional plumage very similar to Temminck's Stint, but always distinguishable by its short, black and straight bill (1) and the greyish-black legs (2). The unmoulted scapulars, coverts and breast feathers from the br plumage are still evident (3). This species is mostly seen in small numbers during a brief period in late summer.

Purple Sandpiper. 15.9. KBJ. Medium-large and stout *Calidris* with characteristic blackish-grey plumage all year round. Among the unmoulted blackish scapulars with rusty-red edges and among the wing coverts, the new greyish-black winter feathers become evident (1). Always distinguishable from Dunlin by the blackish-grey head, the bicoloured bill and the yellow legs.

Sanderling. 7.8. KBJ. During late summer the reddish br plumage is moulted to the characteristic winter plumage with grey upperparts and pure white head and underparts. Seen in small parties sprinting along the tide-line on sandy beaches. Differs from Dunlin by behaviour and the short straight bill (1). The only *Calidris* sandpiper with no hind toe (2).

Temminck's Stint. 15.8. DP. One of the two European stints. During late summer the reddish-beige scapulars are moulted and thereafter it becomes very similar to Little Stint. Always recognisable from the latter species by the up-turned lower mandible (1), which looks black, but is often dark yellowish-brown on the inner half as well as by the greenish-yellow legs (2).

Curlew Sandpiper. 16.8. JL. The only Dunlin-sized wader with brick-red head and underside in the summer season. May resemble the considerably larger Red Knot, but is told from this species, and from all other waders by its curved bill (1). Migrates over land and along the coast.

Red Knot. 1.9. HS. A short-legged, compact and stoutly built *Calidris*, considerably larger than Dunlin. The br plumage is moulted to a drab winter plumage with grey upperparts and pale underside. Seen in flocks on tidal flats, often together with Bar-tailed Godwit which has a similar plumage but is larger with longer bill and legs.

Ruff. ♂. 11.9. HJE. The spectacular breeding plumage with ear-tufts and ruff is moulted after the mating season. The remains are seen on the upperside and belly (1). The ♂ is normally one quarter larger than the ♀ and has a more triangular head shape. In the br season both sexes have orange-yellow legs (2).

Dunlin. 9.8. LG. Most common European *Calidris* sandpiper and a reference species for other *Calidris* species. Similar in body size to a Common Starling. Has medium-long, black and decurved bill (1), medium-long black legs and (in br plumage) is the only *Calidris* species with all-black belly patch (2).

Western Sandpiper. 8.8. DP. One of seven stints. Similar to Dunlin, but is noticeably smaller with shorter and faintly decurved bill (1) as well as distinct, dark chevrons on breast, flanks and vent (2). Western Sandpiper and Semipalmated Sandpiper are the only *Calidris* species with partially webbed toes.

Least Sandpiper. 6.6. DP. Smallest of the seven stints and one of three with yellowish legs. Resting birds look almost spherical. Recognised by large, dark centres to the scapulars (1), narrow and pointed bill (2), yellow legs (3), and tertials with notches along the edge (4), which separates it from Long-toed Stint.

Semipalmated Sandpiper. 28.6. GV. Very small and pale. Similar to Western Sandpiper, but has shorter, almost straight bill (1) and only a suggestion of brown chevrons on upper breast (2). Semipalmated Sandpiper and Western Sandpiper are the only *Calidris* species with partially webbed toes.

Long-toed Stint. 8.6. DP. The only Asian stint with yellowish legs (1). Similar to Least Sandpiper in br plumage, but has paler head and mantle and lacks notches in the cinnamon-coloured edges of the tertials (2). Has long toes like Least Sandpiper, but with longer middle toe (3).

Red-necked Stint. 3.5. DP. One of seven stints, a rare vagrant from Asia. Similar to Little Stint, but differs by having grey wing coverts without dark centres and orange-red edges (1) the grey tertials with narrow pale edges (2). Unlike Little Stint does not have pure white throat (3), and often lacks the latter's pale 'V' on mantle.

Baird's Sandpiper. 4.6. DP. Long-winged with long primary projection. Similar to White-rumped Sandpiper, but differs by having unstreaked flanks (2) and black bill (3). The ground colour of br plumage is mainly buffish-brown on upperside and breast, contrary to the whitish colour of the White-rumped Sandpiper.

White-rumped Sandpiper. 8.6. DP. Smaller than Dunlin. Has long primary projection (1). Differs from the Baird's Sandpiper by a more black-and-white, colder plumage with a faint buffish tinge. Has diagnostic white rump, streaked flanks (2) and slightly decurved bill with yellowish-brown base to lower mandible (3).

Sharp-tailed Sandpiper. 6.5. DP. Dunlin-sized. Rare vagrant from Asia. Resembles the slightly larger Pectoral Sandpiper, but can be distinguished by its chestnut crown (1). Clear white eye-ring (2). Diffuse transition between breast and belly markings (3) with chevrons on the flanks (4).

Pectoral Sandpiper. ♂. 5.6. DP. Larger than Dunlin. Has brownish bill with yellowish base of lower mandible (1). All year round easily told by plain plumage with a diagnostic breast-pattern, the streaks often finishing in a point at middle of the breast, and sharply demarcated against the pale belly (2).

Stilt Sandpiper. 8.6. DP. Body-wise slightly large than Dunlin, but easily told from all other Calidris species by a combination of long yellow legs (1), a long slightly decurved bill with blunt tip (2), rusty-red ear-coverts (3) and sides of crown (4) and the barred underside (5).

Buff-breasted Sandpiper. June. KK. Dunlin-sized *Calidris* sandpiper. Resembles Ruff, but differs by round head with large beady black eye in a pale face (1), short, tapering greyish-black bill (2) and orange-yellow legs (3) which match the beige plumage with large, black-centred feathers on the upperside.

Dunlin. 9.8. LG. A reference species for all other *Calidris* species. Similar in size to Common Starling. Has medium-long, black and slightly decurved bill (1) and medium-long legs. In fresh plumage resembles a toned-down version of the ad in br plumage with diffuse belly patch, formed by blackish spotting (2).

Western Sandpiper. 8.8. DP. Resembles juv Little Stint. Has white 'V' on mantle and rusty-red upper scapulars. Differs from Semipalmated Sandpiper by tapering bill (1), rusty-beige scapulars, the middle-row have black subterminal anchors (2) and short primary projection (3). Has partially webbed toes.

Least Sandpiper. 8.8. DP. One of three stints with yellowish legs (1). Recognised from juv Temminck's Stint and juv Long-toed Stint, which also have yellowish legs, by its Little Stint-like look and long toes respectively (2) and pale supercilium adjoining the base of the bill and separate crown from bill (3).

Semipalmated Sandpiper. 8.8. DP. Resembles a pale version of juv Little Stint with beige, streaked breast markings which may form a breast band (1) as well as dark greyish-green legs (2) and short thick bill with blunt tip (3) separating the species from the similar Western Sandpiper.

Long-toed Stint. 8.6. DP. One of seven stints and one of three with yellowish legs (1). Differs from juv Least Sandpiper in having more contrasting plumage, more upright stance and slender body. Additionally, streaks on crown reach the base of bill (2), and has very long middle toe, longer than the tarsus and longer than on Least Sandpiper.

Red-necked Stint. Post juv. 3.10. DP. One of the seven stints. Rare vagrant from Asia. Has grey winter feathers on mantle and scapulars (1). The row of black-centred scapulars with rusty-red fringes (2) are the only colourful feathers on an otherwise grey upperside, separating the species from juv Little Stint which has more contrasting plumage with white 'V' on mantle.

Baird's Sandpiper. 7.9. HS. Smaller than Dunlin. Long primary projection (1). Differs from juv White-rumped Sandpiper on scaly greyish-brown upperside. Streaked neck and upper breast clearly demarcated from the white belly (2). Unstreaked flanks (3). White 'bridge of nose' (4) and white projecting lower mandible (5).

White-rumped Sandpiper. 11.10. JSH. Smaller than Dunlin. Long primary projection (1) separates the species from juv of the seven stints. Differs from juv Baird's Sandpiper by the lower mandible colour (3), lack of projecting lower mandible (3), streaked flanks (4) and the overall colder tone.

Sharp-tailed Sandpiper. 12.9. GV. Differs from ad br by black-centred, white-edged scapulars (1) and whitish-beige, unbarred underparts. Told from similar juv Pectoral Sandpiper by rusty-red edges of scapulars and tertials (2) pinkish-beige, almost unstreaked breast (3) and by short greyish bill.

Pectoral Sandpiper. 17.10. DP. Slightly larger than Dunlin. Differs from other similar *Calidris* sandpipers by the streaked breast sharply demarcated from white belly (1), colour and shape of bill (2) yellowish legs (3). Told from the similar juv Sharp-tailed Sandpiper by very pale edges to tertials and coverts (4).

Stilt Sandpiper. Post juv. August. KK. Slightly larger than Dunlin. Differ from other *Calidris* sandpipers on shape and length of bill (1) and long yellow legs (2). Told from most *Tringa* sandpipers by lower coverts and tertials which have uniform pale edges (3) where *Tringa* species have notched edges.

Buff-breasted Sandpiper. 22.9. AA. Size as Dunlin. Differs from ad by having blackish-brown feather centres with pale, frayed fringes on upperside. Told from other *Calidris* species, including ♂ and juv Ruff, by round head with large, black beady eye in a pale face (1), short, greyish-black bill (2) and orange-yellow legs (3).

Spotted Redshank. Juv. 23.8. JL. Has white wedge on back like most larger *Tringa* sandpipers (1). Resembles Common Redshank, but slightly larger and more elegant. Has long, dark thin bill with red inner part of lower mandible (2). Legs are long (3) and project beyond tail, but are often drawn up under belly.

Common Redshank. Ad br. 20.5. BLC. Most common *Tringa* and a reference species for other species in the genus. Distinguished from most by the red, medium-long legs (1). The broad white wing patches (2), and bicoloured bill, red at base and black towards tip (3). Juv has greyish bill and orange legs.

Common Greenshank. Juv. 16.8. JL. Largest European *Tringa*, larger than Common Redshank and with medium-long and stout bill (1), which is slightly upturned, grey and black towards tip. Legs greenish-yellow (2); pale greyish-green in winter. The greater part of the toes reach beyond the tail.

Marsh Sandpiper. Ad br. 21.4. HS. Slightly smaller than Common Redshank and resembles a more elegant and slender version of Common Greenshank. The size and needle-thin bill (1) sets it apart from other *Tringa* species throughout the year. Legs are yellow to greenish-yellow, and feet reach beyond the tail (2).

Greater Yellowlegs. 25.1. KK. Rare vagrant from North America. Does not have white back-wedge like most larger *Tringa* sandpipers. Resembles Common Greenshank with slightly upturned and finer bill, greyish at base and dark at tip (1). Has long yellow legs (2) and looks like Lesser Yellowlegs, from which the size difference is key.

Lesser Yellowlegs. Ad. 8.8. KK. Rare vagrant from North America. Resembles Wood Sandpiper but is larger, with medium-long, slender dark bill (1) and conspicuous medium-long, yellow legs (2). Distinguished from Greater Yellowlegs on smaller body, shorter, straighter bill and shorter yellow legs.

Green Sandpiper. 10.7. NJL. Small *Tringa* with brownish-black spotted upperparts and white body. Similar to Wood Sandpiper, but has fewer (3–4) broad black bars on tail (1). Greyish-blue to greyish-green legs (2), black hand lacking white shaft on outer primary (3). Blackish-brown underside of wing with barred axillaries.

Wood Sandpiper. Ad br. 15.7. JL. Small *Tringa* resembling Green Sandpiper, but differs in having greyish-brown upperside with larger pale markings. Yellowish-green legs with feet projecting beyond tail (1), finer barring on tail (2), greyish-white underwings with brownish barring (3), white shaft on outer primary (4).

Common Sandpiper. Juv. 3.8. JL. Recognised by its flight with stiff and shallow wing-beats, accompanied by the excitable call '*hee-dee-dee-dee*'. Has grey bill, which in the br season may have dark yellow base (1). Long rounded brown tail (2) and distinct white wing-bar almost reaching the body (3).

Solitary Sandpiper. Juv. 30.8. TH. Rare vagrant from North America. Very similar to Green Sandpiper, but is a bit smaller and more slender with slightly narrower and longer wings. Primarily distinguished from Green Sandpiper by the evenly barred tail with black centre (1).

Terek Sandpiper. 23.3. MYJ. The same size as Green Sandpiper. Separated from other waders by its bicoloured, upturned bill (1) and orange-yellow legs (2). Upperside of tail is uniform grey, the wings are white below and brown above with white trailing edge (3) on the arm and half the hand.

Spotted Sandpiper. Ad in br plumage. 24.6. KK. Rare vagrant from North America. In summer told from Common Sandpiper by reddish-yellow bill, which is brownish in the winter season (1), and characteristic, distinct spots on underside (2) though some individuals have weaker spotting. Also, wing-bars do not reach the body (3), and has shorter tail (4).

Spotted Redshank. 6.5. BLC. Large, elegant *Tringa* with long legs and long thin bill. Unmistakable in its black br plumage, where even the legs are black, on occasion partly red (1). The rest of the year legs are bright red. Differs from other *Tringa* species by black bill with red base of lower mandible (2).

Common Redshank. 6.5.LG. Europe's commonest *Tringa* is medium-large and a reference species for the genus. Recognised by red legs (1). Outside the br season differ from Spotted Redshank by bicoloured bill, red on inner half, black on outer (2). The subspecies *robusta* is greyish with black chevrons on underside.

Common Greenshank. 4.7. DP. The largest *Tringa* in Europe. Distinguished all year round by size and the characteristic, stout and slightly upturned bill, which is grey on the inner half and black on the outer (1). The legs are greenish when br, greyish to greysh-green the remainder of the year (2).

Marsh Sandpiper. 6.5. DP. Smaller than Common Redshank and resembles a smaller and more dainty version of Common Greenshank. Recognisable all year round by its pale appearance, the rather long needle-thin straight bill (1) and the very long greenish-yellow legs (2).

Greater Yellowlegs. 4.4. KK. Rare vagrant from North America. Resembles Common Greenshank, but differs in having yellow legs (1), upperside mottled in black (2) and in the barred underside (3). Told from the similar Lesser Yellowlegs by size and the longer, more robust bill, which has greyish to greenish-yellow basal part (4).

Lesser Yellowlegs. 3.4. KK. Rare vagrant from North America. Resembles a large Wood Sandpiper, but has longer and yellow to yellow-orange legs (1) which sets it apart from other *Tringa* species of the same size. Bill is medium-long and slender with faint yellowish base (2). Differs from Greater Yellowlegs on size and shorter legs and bill.

Green Sandpiper. 2.7. LG. Small and elegant. Appears all black-and-white, but at close range the white flecks on the brownish-black upperside (1), dense streaks on neck (2) and the greyish-green to greyish-blue legs become evident. Similar to Wood Sandpiper, but differs in longer bill with a curved tip (3).

Solitary Sandpiper. May. KK. Rare North American vagrant. Slightly smaller and more slender than Green Sandpiper. Has mostly broader, more conspicuous white eye-ring (1). Another confusion species is Wood Sandpiper, but Solitary has jizz and plumage more like that of Green Sandpiper.

Wood Sandpiper. 4.7. LG. Small *Tringa* species. Similar to Green Sandpiper, but has greyish-brown upperside with larger pale spots and flecking (1) shorter bill, with yellowish-brown on inner part (2). Greyish-white underside with diffuse brown streaks on neck, breast and barred flanks (3), and yellowish-green legs (4).

Terek Sandpiper. 27.5. HS. Unmistakable smallish grey-and-white wader with long thin upturned bill (1) and orange legs (2). In the br season has conspicuous broad black 'shoulder-straps' (3) and black shaft streaks on scapulars and wing coverts (3). In addition it has webbed toes.

Common Sandpiper. 11.6. JL. The smallest, shortest-legged and longest-tailed *Tringa* sandpiper (1). Differs from other species by having a white wedge between the greyish-brown upper breast and the carpal joint (2). Often has yellowish bill during br season, particularly on the inner half (3), whereas it is greyish-black the remainder of the year.

Spotted Sandpiper. May. KK. Rare vagrant from North America. In the br season distinguished from Common Sandpiper by reddish-yellow bill (1), coarsely spotted underside (2), albeit ill-defined on some individuals, reddish legs (3) and a shorter distance from the tip of longest primary to the tip of the tail than on Common Sandpiper (4).

Spotted Redshank. 31.7. JL. Large and elegant *Tringa*. Recognised by size and shape, and from ad by the dusky, mottled plumage and orange legs (1). Told from other *Tringa* species, by the blackish bill with orange-red inner part of the lower mandible (2). Unlike ad at this time of year, the juv are in fresh plumage.

Common Redshank. 20.7. HS. The most common *Tringa* in Europe. Medium-sized and a reference species for the genus. Resembles juv Spotted Redshank, but has smaller body, shorter bill (1), which is yellowish-brown on inner half, shorter yellow-orange legs (2) and pale, streaked belly (3).

Common Greenshank. 19.8. LG. The largest of the genus in Europe. Recognisable by size and the stout and slightly upturned bill, which is grey on inner half and black on outer (1). Differs from ad in having fresh plumage with pale fringes on the upperside and in late summer by the green legs (2).

Marsh Sandpiper. 19.8. DP. Smaller than Common Redshank and resembles a Common Greenshank. Recognisable by its pale appearance and the medium-long, straight and needle thin bill (1) long greenish-yellow legs (2). Differs from ad in late summer by having fresh plumage with broad pale edges on upperparts.

Greater Yellowlegs. 9.8. KK. Rare vagrant from North America. Resembles juv Common Greenshank, but differs by having long yellow legs (1) and pale spotted upperside (2). Unlike the ad, juv have fresh plumages at this time of the year. Recognised from Lesser Yellowlegs by larger body, longer bill (3) and longer legs.

Lesser Yellowlegs. 2.9. TH. Rare vagrant from North America. With respect to plumage and shape of body, it resembles juv Wood Sandpiper, but is larger, possesses longer yellow legs (1) and a longer bill (2). Differs from ad Greater Yellowlegs by smaller size, shorter bill and legs.

Green Sandpiper. 7.8. HS. Appears black-and-white, but at close range the white flecks on the brownish-black upperside (1) become visible. Differs from ad with fresh plumage and washed-out streaks on neck (2). Resembles juv Wood Sandpiper, but has longer bill (3), white belly and flanks (4) greyish-green legs (5).

Wood Sandpiper. 9.8. LG. Small *Tringa* species. Similar to juv Green Sandpiper, but recognisable by brown upperside with pale edges to mantle-feathers and serrated coverts and tertials (1). Has shorter bill (2), greyish-white underside with diffuse brown streaks on neck, breast and barred flanks (3), yellowish-green legs (4).

Common Sandpiper. 9.8. LG. The smallest, shortest-legged and longest-tailed *Tringa* sandpiper (1). Differs from other species in having a white wedge between the greyish-brown upper breast and carpal joint (2). Told from ad by fresh plumage with brownish-beige fringes, and dark notching on the edges of tertials (3).

Solitary Sandpiper. 30.8. TH. Rare North American vagrant. Slightly smaller and more slender than the nearly identical Green Sandpiper, with broader, more conspicuous white eye-ring (1). The spots on the upperside are larger (2). Another candidate for confusion is juv Wood Sandpiper.

Terek Sandpiper. 27.8. DP. Smallish grey-and-white wader with diagnostic long thin upturned bill (1) and short orange legs (2). Differs from ad in br plumage in lacking black 'shoulder-straps' and black shaft streaks on scapulars and wing coverts (3). The Terek Sandpiper has webbed toes.

Spotted Sandpiper. Post juv. 11.11. KK. Rare vagrant from North America. Distinguished from juv Common Sandpiper on greyish upperside with buff fringes (1) and broad white-tipped tertials with dark subterminal bars (2). Juv Common Sandpiper has dark notching along edge of tertials and U-shaped subterminal bars.

Eurasian Woodcock. 23.12. JLA. The only species of woodcock in Europe is medium-large and short-legged. Told from the snipes by bulky and deep-chested body, triangular head with golden barring on the dark crown (1), and densely barred underside (2). Breeds in forests and is mainly nocturnal.

Eurasian Woodcock. 18.10. SEJ. In flight identified from the snipes by broad, arched wings and plump, barred body blending in with the barred underwing (1). The snipes have paler bodies and more contrasting undersides of wings.

Great Snipe. 13.6. DP. Large and heavy snipe, in stature not unlike Eurasian Woodcock. Differs from the other medium-sized snipes by shorter bill (1), distinct chevrons on the flanks (2), white outer tail feathers (3), and the rows of white bands on wings formed by white-tipped greater and median coverts (4).

Great Snipe. 27.7. JL. Plump and heavy snipe. Larger than Common Snipe, from which it can be identified by shorter bill (1), distinct chevrons on flanks (2), and evenly grey underwing with black-and-white V-shaped markings on the axillaries (3). Furthermore, the upperside of the wing shows three well-defined wing-bars.

Jack Snipe. 26.10. HJE. The smallest snipe. Body size same as Common Starling, and with relatively short bill (1). Inconspicuous behaviour and thus rarely seen. Recognisable by size and the eye-catching broad, golden stripes along back and scapulars (2). Feeds with repeated swinging movements of front body.

Jack Snipe. 21.11. JL. The smallest species of snipe; body size same as Common Starling. Resembles Common Snipe in flight, but can be told from it by shorter bill (1) and broad, golden stripes on the back (2). The toes do not protrude beyond the short tail (3) and the silhouette can give a bat-like impression.

Common Snipe. 20.5. JL. The most common snipe in Europe. Recognised from other European snipes by the very long bill (1). The toes project just beyond the tip of the medium-long tail (2). Distinct white trailing edge to the arm (3). Barred underwings with broader white barring than black on the axillaries (4).

Common Snipe. 9.8. LG. The most common snipe in Europe. Medium-sized with rather long yellowish-green legs (1) and a very long bill at up to 7 cm, or approximately twice the length of the head (2). The outer webs of the scapulars have straw-coloured edges (3), while the inner edges are reddish-brown.

Pintail Snipe. 6.3. HS. Rare vagrant from Siberia. Resembles Common Snipe, but has larger eyes (1) and a shorter tail (2). Toes project well beyond the tail (3). Boldly barred underwing where the axillaries have broader dark barring than white (4). Has a finer white trailing edge to the arm (5) than Common Snipe.

Pintail Snipe. 17.9. AJ. Rare vagrant from Siberia. Resembles Common Snipe, but has larger eyes (recalling Eurasian Woodcock) (1) and short tail (2). The scapulars are edged yellowish-beige on both sides (3) whereas Common Snipe has reddish-brown and whitish-yellow edges on respectively inner and outer edges.

Wilson's Snipe. 23.12. BS. Rare vagrant from North America. Almost identical to Common Snipe. Differs in having very narrow or no whitish trailing edge of arm (1), and on uniform grey underwing where the axillaries have broader dark barring than pale (2). The white barring on these feathers is broadest on Common Snipe.

Wilson's Snipe. 23.12. BS. Rare vagrant from North America. Almost identical to Common Snipe. The plumage varies from dark to pale, with or without densely barred flanks. Almost impossible to tell resting birds from Common Snipe without additional characters of upper-and-underwing.

Red-necked Phalarope. ♀. 23.5. JL. The size of Dunlin and unmistakable in br plumage. Differs from the less colourful ♂ by having deeper orange-red colour on neck (1) and on generally darker, blackish-grey body plumage. Always recognisable from Grey Phalarope by the needle-thin black bill (2).

Red-necked Phalarope. ♂. 23.5. JL. Unmistakable in br plumage. Differs from the ♀ in having more subdued greyish-black body plumage, also paler orange-red markings on the neck (1), and buff edges to the larger scapulars (2). Always recognisable from Grey Phalarope by the needle-thin black bill.

Red-necked Phalarope. Juv. 1.9. LG. Recognised from juv Grey Phalarope by the needle-thin bill (1) and the warm buff stripes on mantle (3). Overall more slender and elegant in appearance without Grey Phalarope's deep chest and tendency to adopt hunch-backed posture.

Grey Phalarope. ♀. 24.6. HS. The size of Dunlin. Unmistakable in br plumage. Differs from the ♂ on all-black crown (1) and black-and-white upperside without golden edges on the scapulars. Differs all year round from Red-necked Phalarope on the broader and slightly shorter bill (3) as well as the more robust shape of body with more rounded back and deeper chest.

Grey Phalarope. ♂. 6.6. DP. The size of Dunlin. Unmistakable in br plumage. Differs from the ♀ by the mottled brownish crown (1), and on the golden edges to mantle and scapulars (2). Notice the toes (3) which are lobed and partially webbed. This enables the Grey Phalarope to swim and stir up food from the bottom. The shape of the lobes differs between the three species.

Grey Phalarope. Juv. 15.10. EFH. Dunlin-sized. Differs from juv Red-necked Phalarope by heavier body with more rounded back, thicker neck and deeper chest, also by the shorter and heavier bill (1). More or less brownish-beige markings on neck and breast (2). Grey back (3) without golden, and later in winter, white stripes.

Wilson's Phalarope. ♀. 4.6. DP. Unmistakable in br plumage. The largest phalarope. Body size not that much larger than the other two species, but the legs and bill are noticeably longer. The ♀ has brighter and more well defined colours than the ♂.

Wilson's Phalarope. ♂. 4.6. DP. Colourwise, very subdued and thus markedly different to the colourful ♀. Told from the two smaller species by longer legs and bill. The species feeds on land more frequently than Red-necked and Grey Phalaropes.

Wilson's Phalarope. Post juv. 12.8. DP. Has a few newly emerged grey first-winter feathers on mantle and scapulars (1). Easily told from ad and from the other phalaropes by the medium-long yellow legs (2). Notice the not very clearly webbed and lobed toes.

Red-necked Phalarope. Juv. 28.8 HS. Size of a Dunlin. As a rule migrates singly or a few together, but is often seen in mixed parties of *Calidris* sandpipers. Identified by the needle-thin bill (1) and the golden stripes on the back (2). Distinguished in flight from juv Grey Phalarope by the bill and the overall more elegant and slender shape, lacking deep chest.

Red-necked Phalarope. Ad moulting to non-br. 10.11. ISA. Identified by lack of golden edges on scapulars and tertials, also by the contrast between the faded and worn tail feathers, flight feathers and tertials, and the newly grown, greyish-blue scapulars and mantle feathers broadly edged whitish. Always distinguishable from Grey Phalarope by longer and needle-thin bill (2).

Red-necked Phalarope. Ad non-br. China. 3.10. DP. Identified by the nearly white head and greyish-blue upperparts with white-striped mantle (1), broad white edges to scapulars (2), and fresh tertials with thin white edges (3). The tip of a worn primary (4) is just visible above the white, frayed undertail coverts. Told from Grey Phalarope by the thin bill.

Grey Phalarope. Juv. 27.9. SEJ. Size of Dunlin. Distinguished in flight from juv Red-necked Phalarope by thicker and shorter bill, which may have yellowish base (1). All-grey back without golden stripes (2), and a more broad-winged and muscular jizz with deep pigeon-like breast. Often seen in the surf, launching itself fearlessly into the waves in pursuit of morsels of food.

Grey Phalarope. Post juv moulting to first winter. 20.9. HS. Identified by the blackish-brown remains of juv head markings (1) and new grey feathers on mantle and scapulars with a few unmoulted black, beige-edged feathers (2), and the pale-edged juv tertials (3). Told from Red-necked Phalarope by the stout and shorter bill, which may show a yellowish base even when the bird is in juv plumage (4).

Grey Phalarope. First winter. 11.12. NLJ. Identified by the juv, blackish-brown tertials with narrow whitish to beige edges (1). Ad non-br has grey tertials. Grey Phalarope always told from Red-necked Phalarope by the relatively sturdy bill with yellow base (2), and in the winter season by the uniform, greyish-blue mantle without whitish stripes (3).

Wilson's Phalarope. Juv. 25.8. KK. Contrary to the European phalaropes, this species has a plain upperside with white rump and uppertail coverts, though covered by the tertials on this photo (1). Very thin white wing-bar along the greater coverts (2). Additionally, juv have pale-edged median coverts (3).

Wilson's Phalarope. First winter. 28.9. RSN. Identified by dark grey crown and the white supercilium (1), the narrow black eye-stripe (2), the long, needle-thin bill (3), and the long neck and attenuated shape of body. Aged by the dark tertials with narrow pale edges (4).

Wilson's Phalarope. Ad br to non-br. 7.7. BS. On the water it resembles a long-necked and elongated Red-necked Phalarope, but in all plumages differs by the longer bill (1). The non-br plumage is very pale grey and white with a thin grey stripe behind the eye extending down the neck.

Dunlins, ad br. Bar-tailed Godwit. 5.5 LG.

Dunlins, ad br to non-br and juv. 29.8. LG.

Sanderlings, ad br. Dunlins; ad br. 31.7. KBJ.

Ringed Plovers, Broad-billed Sandpipers and Dunlins. All ad br. 10.5. DP.

Curlew Sandpipers, ad br. 16.7. NLJ.

Ruddy Turnstones, Dunlins and Ringed Plovers, ad br. Red Knot, immature still in non-br plumage. 24.5. JL.

Purple Sandpipers, non-br to br. 17.4. JL.

Red Knots, non-br. 26.1. JL.

Grey Plovers, ad br and juv. 9.8. AK.

Ruffs, ad br. 24.4. NLJ.

Common Greenshanks, juv. Curlew Sandpipers, ad br to non-br and juv. Red Knot, juv. 17.8. JL.

European Golden Plovers, ad br and non-br to br. 15.4. JL.

Common Redshanks. 9.8. LG.

Eurasian Oystercatchers. Eurasian Spoonbills in the background. 11.8. LG.

Marsh Sandpipers. 10.7. Greece. NLJ.

Common Snipes. 18.9. JL.

Bar-tailed Godwits, ad br ♂ and ♂. At the top right Red Knot, ad br. 26.8. Sweden. NLJ.

Collared Pratincoles, ad br. 27.4. Andalusia, Spain.

Dunlins, ad br to non-br with juv at roosting place in the Wadden Sea. 28.9. LG.

Dunlins, ad br. Little Stint, ad br. 21.7. HS.

Broad-billed Sandpiper, juv. Dunlin, juv. 4.9. HS.

Ruddy Turnstone, Lesser Sand Plovers and Curlew Sandpipers, ad br with Lesser Sand Plovers, non-br. 22.3. Khok Kham, Thailand. HS.

Common Ringed Plovers, ad and juv. Dunlins, post juv. 26.9. HS.

Sanderlings, engaged in typical feeding-fashion along the surf. 27.5. LG.

Red-necked Phalaropes, non-br. 6.11. The Indian Ocean east of Oman. ISA.

Common Redshanks (*T. t. robusta*), Ruddy Turnstones and Dunlins. All non-br. 31.1.BLC.

Purple Sandpipers, non-br. Varanger, Norway.12.3. HS.

European Golden Plovers, ad br. 6.5. HS.

Bar-tailed Godwits, ad br, ♂ and ♂. Red Knot, ad br. Dunlin, juv. Eurasian Oystercatchers. Also Black-headed Gull and Sandwich Tern. 26.8. HS.

Black-tailed Godwits, *L. l. islandica* ad ♂ and ♂, non-br to br. Wash Bay, England. 7.4. JKAM.

Waders

This bird shows long legs and a long bill, so it must be some kind of wader.

But which kind?

Compared with those rocks, it is a small one.

A *Calidris* sandpiper maybe?

No, it seems a bit too robust around the chest, and it has a very long tail.

Maybe it's a Ruddy Turnstone? That fits in well with the habitat.

But the Ruddy Turnstone is shorter-legged with a shorter and heavier bill.

It has a white wedge between the breast and the front of the wing, and it bobs its tail, just like a wagtail.

It's a Common Sandpiper, of course!

Learn more about the Common Sandpiper; silhouettes, field identification, calls, breeding biology and much more, on the following spreads where 82 common as well as rare species of waders in the Northern Hemisphere are described in detail.

A Common Sandpiper, showing its typical long-tailed and broad-breasted silhouette.

Not seen in any other European species of wader, the white wedge between breast and carpal joint is a diagnostic field mark for the Common Sandpiper. 2.7. LG.

The Eurasian Oystercatcher is one of the most conspicuous waders in Europe. The characteristic black-and-white plumage and the long red bill are highly visible and recognisable from afar.

The Eurasian Oystercatcher occurs within an area extending from Iceland and the western European shores eastward to Russia and Asia, the Siberian grass steppes and the Pacific coasts of Russia and China.

The oystercatcher family numbers ten species, between them occupying most of the coastal areas of the world. All have brightly coloured legs and bill and most are black-and white, but there are also all-black species in America, Africa and along the coasts of Australia and New Zealand.

Eurasian Oystercatchers take off from a tidal roosting place in the Wadden Sea. 10.04. LG.

EURASIAN OYSTERCATCHER
HAEMATOPUS OSTRALEGUS

Meaning of the name
'The blood-footed oyster gatherer'
Greek: *haimo*, blood and *pous*, foot.
Ostreion, oyster, *lego*, gather.

Jizz
L. 40–45 cm. Bill, male 7.6 cm, female 8.1 cm. Ws. approx. 77 cm.

Unmistakable among waders by its bold black-and-white plumage and the long, straight red bill. When standing it displays a solid, broad-chested, pigeon-like body. Very noisy in the breeding area. Often seen in large flocks in winter.

In flight. The broad white wing-bars and the wedge of white on the back are striking. The flight is straight, often close to the surface of the water, with stiff strokes of the slightly low-hanging wings.

▲ Adult breeding. Male in the breeding area proclaiming his territory with the characteristic 'cubeek, cubeek, cubeek' call. Note the totally black back, the all-red bill and the broad black band on the tail. 12.6. LG.

▲ Juvenile. Brownish-black, vermiculated mantle and scapulars with beige fringes and ill-defined white chin-band distinguish the juvenile from the first-summer bird, which has black mantle and scapulars, brown coverts and a more distinct white chin-band. 19.8. JL.

Plumage and identification

Eurasian Oystercatchers seem fairly similar throughout the year, yet have several distinctive features that enable us to distinguish between four different age groups, ranging from juvenile to adult first-time breeders, which are three-year-olds (third summer). The Eurasian Oystercatcher has the longest immaturity period among waders.

Adult breeding, male. Bill is orange year-round, and is slightly shorter and more robust than that of the female. Dark red eye with a thickened orbital ring and a black-and-white plumage with broad white wing-bars, a white wedge on the back and white tail with a broad black terminal band. The lower breast, belly and vent are white and the legs dark pink.

Female. Like the male but with a longer and a narrower bill and a brownish tinge to the upperside.

Adult non-breeding. Like breeding, but with a white chin-band, often narrower and more blurred than in the juveniles. The white chin-band is acquired during late summer and lost in early winter. A small percentage of the adult birds lack this chin-band.

Juvenile. Has a dark eye with a dark, narrow orbital ring, and orange bill with dark culmen and tip. A hint of indistinctly marked chin-band may be seen. The mantle and scapulars are brown-tinged and vermiculated. Legs grey.

First winter. Moulting into all-black back feathers and developing a narrow white chin-band.

Immature. Young birds until the third summer (when fully adult) may be distinguished by an increasingly red iris, a broader and more intensely coloured orbital ring and a more vividly orange bill with a yellowish tip as well as pink legs.

Subspecies

Besides the nominate form, *H. o. ostralegus*, the subspecies *H. o. longipes* occurs in the north-eastern Mediterranean, the Adriatic and eastwards as far as Russia and western Siberia. This subspecies has a longer bill and legs, a longer and deeper nasal groove and blackish-brown mantle and scapulars. In the field, the identification may be difficult owing to the existence of intermediary stages.

▲ A first summer individual with white chin-band, reddish-brown iris and dusky bill. Note that the bird is moulting into the second winter with new primaries and shows conspicuous moulting gap, caused by a mixture of growing new inner primaries and old unmoulted outer ones. 2.6. LG.

▼ Adult breeding male with black back and a shorter and more robust bill than that of the female, which has brownish-tinged back feathers. Only adults show a bright orange bill with yellow tip. 6.7. LG.

▲ The Asian subspecies *H. o. longipes*, second year non-breeding, has longer legs and bill than the nominate and also a blackish-brown upperside. In female *longipes*, the bill may be as much as 9 cm long. January, Oman. HJE.

During the six months of winter, when the sand-burrowing molluscs withdraw further into the seabed, the Eurasian Oystercatchers are reduced to hard food, and the tip of the bill is worn into a chisel-like shape by hammering through thick mussel shells.

During the summer season, when the breeding pair moves into their territory and the available food in the shape of worms and insects is now softer, the bill grows as much as one centimetre and assumes a pincer-like shape.

Note that the pupil appears slightly blurred in the male to the left, due to an adjacent dark eye-spot.

In the Black Oystercatcher (*H. bachmani*) of North America, researchers have correctly determined the sex of 94% of a test group by applying the theory that females have a full eye-spot and males a slightly blurred one or none at all.

Whether this finding can be applied to the Eurasian Oystercatcher is uncertain.

The cause of the matter, which is incidentally also found in Black Woodpecker and Woodpigeon, where it is not related to sex, remains uncertain.

Voice

The Eurasian Oystercatcher has a loud voice which can be heard from afar at the onset of the breeding season, when the territory needs to be occupied and defended. The most frequently employed calls are the alarm call, an indignant, shrill 'bjeek-bjeek-bjeek' and the contact call which is a loud 'cabeek-cabeek', which ends in a dimishing trill.

In addition, a characteristic song-flight is performed above the territory.

The nest may be a depression in the pebbles on the beach or, as here, a scraped-off patch in short turf, lined and decorated with sea shells. The first clutch usually consists of three eggs, one fewer than the normal number for most waders.

If nests are predated, for example by foxes or gulls, or lost because of flooding, the female may, like other wader species, lay as many as three replacement clutches if her condition permits it. But the process is energy-demanding, and results in smaller clutches with each successive laying. The breeding attempt is often given up completely. 26.5. LG.

Habitat

Breeds along sandy, gravelly and rocky coasts, on meadows and in marshland along the coast. In the north-western part of its distribution also on cultivated fields, along rivers and by larger lakes, gravel pits and on moors with bare ground and stretches of meadowland.

Outside the breeding season the species is primarily seen on the coast, often in larger flocks in protected bays and at estuaries.

Breeding biology

Usually breeds from the age of four or later. The pair remains together for life and individual birds may live for 40 years or even more.

Both adults defend the territory, both at the nesting ground itself and in the adjacent foraging areas. Three to five, most often three, eggs are laid in one clutch at some time between mid April and the beginning of June. Incubation is by both sexes for 24–28 days.

The chicks leave the nest as soon as they are dry, keenly watched by the parents, which on an average only manage to raise one young to fledging age, often none at all, on account of heavy predation by Herring and Great Black-backed Gulls. In actual fact though, the young are given the very best start in life, as the Oystercatcher is one of the few waders that initially feeds its offspring. After a minimum of 33 days the young can fly and in late summer they accompany their parents on migration towards the wintering areas, where they join large flocks in food-rich regions like the Wadden Sea.

Migration

The birds from northern Europe primarily winter in the Wadden Sea, with a few along the coast of north-western Africa. Up to 20,00, mainly Norwegian birds, winter in the Danish part of the Wadden Sea and another 50,000 Nordic and Russian birds pass through during their autumn migration. Subspecies *longipes* winters along the east African coast and eastwards to India.

Distribution

The European population is estimated at between 284,000 and 384,000 pairs. The nominate form, *H. o. ostralegus*, breeds primarily along coasts, in certain places also inland, from Iceland, the British Isles and east through Scandinavia to north-western Russia and southwards to north-western France, with isolated populations around the Mediterranean and the Adriatic as far as Turkey.

The subspecies *H. o. longipes* is found in eastern and southern Russia and eastwards to western Siberia.

Asian birds further east belong to the subspecies *H. o. osculans*.

▶ Oystercatchers struggle fiercely for the best territories. Some pairs fly up 20 kilometres from their inland refuges in order to fetch food from better foraging areas than those afforded by the breeding territory. This includes those nesting on lake islands and on buildings with gravel-covered roofs, where they are safe from cats and foxes.

The signalling colour of the bill plays an important part in courtship, territorial defence and in parental care. By pecking at the parent's bill, the two half-grown young trigger the feeding instinct. The reward is regurgitated earth-worms.

The difference in size between the two young shown here is due to the interval of one day and night between their hatching, and presumably more successful begging on the part of the larger chick. 11.6. LG.

▼ Eurasian Oystercatchers in flight. At the rear two adults, non-breeding. Note the fully coloured bill with its intense orange glow and yellow tip, as well as the indistinct chin-band which in adults is moulted in early winter.

At the top in the middle: non-breeding, presumably 3 cy, still with a dark-tipped bill. In the middle (furthest away and slightly out of focus): first winter, with brownish mantle and scapulars, white chin-band, a dark eye and a lighter coloured bill with darker tip than that of the other birds. Of the three birds at the bottom, the one in front without a chin-band is in full breeding plumage. 29.1. JL.

The avocet family consists of two genera, both represented in Europe, namely avocets (*Recurvirostra*) and stilts (*Himantopus*).

The Pied Avocet's dapper plumage and peculiar bill make it unmistakeable among the waders in Europe. Denmark and southern Sweden mark its northern limit.

There are three other avocet species, occurring in North America, South America and Australia respectively.

Up to 6,000 Pied Avocets gather in the Danish part of the Wadden Sea in late summer and autumn, when adult birds undergo their total moult. The birds come from Danish, Swedish and Estonian populations as well as from the adjacent German and Dutch breeding areas.

After completing their moult, the avocets migrate to the wintering areas in the Dutch part of the Wadden Sea and further south along the Atlantic coast. Rømødæmningen 18.9. BLC

PIED AVOCET
RECURVIROSTRA AVOCETTA

Meaning of the name
'The one with the backward-curved bill'.
From *recurvus*, backward-curved and *rostra*, bill.

Possibly the species name of Avocetta refers to the barrister's black and white official garb, but is of uncertain origin.

Jizz
L. 42–45 cm. Bill 8 cm. Ws. *c.* 70 cm.

Unmistakeable large, black and white very long-legged wader with long, thin, upturned bill.

Feeds with swift sideways movements in the surface of the water. Occasionally swims and makes plunges with tail turned upwards like a duck.

In flight. The black and white wing markings flash in flight and draw attention from far away. Flight elegant and swift, often interspersed with gliding. Legs protrude well behind the tail.

Plumage and identification

Pied Avocets have broadly similar plumage all year, but it is possible to distinguish males, females and juveniles.

Adult breeding, male. White with black bill, black to brownish-black crown and nape, black shoulder feathers, black median and lesser wing coverts and black primaries. Legs greyish-blue with partially webbed toes. In spring often has greyish inner wing, fading into white during early summer.

Female. Like male, but slightly smaller and with shorter and more steeply upturned bill. Males and females are easier to tell apart if seen together when breeding. Occasionally shows scattered white feathers in the black crown.

Juvenile. Basically similar to adult plumage, but seems more 'dirty' because of sooty, less clear-cut dark pattern.

First summer. Discernible from adult by retained, very worn juvenile remiges.

Voice

The Pied Avocet is only highly vocal on the breeding grounds, during courtship and territorial defence of eggs and young.

The voice is a variation on the soft '*klytt-klytt*' call, either as a more melodious, song-like phrase or as a more agitated, abrupt or single-note, drawn-out version.

Habitat

In the breeding season, mostly seen on coastal meadows and marshy areas, on islets as well as on freshwater floodplains adjacent to lakes, brackish lagoons and inlets. Also on sandy beaches and salty lagoons.

In winter flocks are seen at sandy and muddy coasts, saltpans and estuaries.

◀ Adult breeding.
Pair of Pied Avocets on marshland, the preferred breeding habitat. 14.4. LG.

▲ Adult breeding. Presumably a female with relatively short and upturned bill. 5.5. JL.

Breeding biology

First breeding occurs at age two or three. Ringing recoveries have shown that adult birds may survive until at least 20 years old.

Breeds in loose colonies with fairly wide distances between nests. The colony may vary in size from a few to several hundred pairs, which jointly defend the nests against gulls, crows and mammal predators, primarily foxes.

The four eggs are laid in a nest lined with dead grass or seaweed and are incubated for approximately 24 days by both sexes. Shortly after hatching, parents and young leave the breeding territory for places with better feeding opportunities. There a feeding territory is maintained until the young are able to fly, after a good 40 days. The young feed independently on an initial diet of insects, small worms and crustaceans until the bill is fully grown and larger worms can be taken in the muddy soil. When fully able to fly, the young join the parents in moving to communal feeding areas, where the family bond gradually dissolves.

Migration

European Pied Avocets arrive at the breeding grounds from March to May and breed from April to July. Before departing from the breeding area, which takes place from August to October, both adults and juveniles gather in traditional moulting localities in the Wadden Sea. The majority of the population winter in the Dutch Wadden Sea and along the Atlantic coast, primarily at estuaries on French, Spanish and Portuguese coasts. In addition, a number of Pied Avocets winter along the north-west African coast.

▲ Adult breeding.

On a cold spring morning, with frost still making the grass crisp, there is hectic activity in the Pied Avocet colony – even before sunrise. They give noisy '*klytt-klytt*' calls in numerous variations, while feints, pursuits and aerial duels are the tactics when rivals are to be chased off. The male with the gently upturned bill dominates from the top of the tussock, while the female with her more sharply upturned bill watches from the left.

Before copulation the pair can be observed in a silent courtship dance with leg bends and simultaneous, backward-turning movements. 14.4. LG

▼ Adult breeding.

Pied Avocets nesting in coastal areas with no good protection from the sea may suffer hardship.

One breeding pair on the Wadden island of Rømø were surprised by a sudden spring tide combined with heavy rains, which flooded the breeding area. The next day, the nest was deserted and emptied by Carrion Crows. This phenomenon is not unknown in tidal areas. The Pied Avocet and other waders, such as Eurasian Oystercatcher, are known to re-lay up to three times, although producing smaller clutches each time, depending on the physical condition of the female. 26.5. LG

▼ Eggs of waders are flecked and spotted with variable ground colour to match the nesting material and the nearest surroundings. The eggs of the Pied Avocet are large, almost the size of hen's eggs. Unfortunately these eggs from this nest to the left never hatched. Notice the wet bottom of the nest. 29.5. LG

Distribution

There are an estimated 58,000 to 74,000 breeding pairs in Europe. In Europe the breeding distribution extends from western European coasts to the east and south, patchily in southern Europe through eastern Europe and southern Russia onwards through southern central Asia to north-eastern China.

The Pied Avocet is also resident in eastern and southern Africa and in the countries adjoining the Arabian Gulf and in north-western India and Afghanistan.

Birds from eastern populations winter from east Africa through the Middle East to India and Myanmar as well as south-eastern China.

▲ Pullus. The downy chick can feed independently from its first hour. The down insulates well and the chick's legs are long and powerful with webbed toes. The bill, though not yet long, is useful for catching insects and for digging in the soft mudflats for worms, sandhoppers and small snails. 13.6. LG.

▶ A couple of juvenile Pied Avocets, in characteristic grubby-looking plumage, in late summer when the bill is nearly full-grown. They can now begin using the feeding technique of the adults, which is to skim the muddy surface for food particles with fast sideways movements.

Note that juveniles roosting on sandbanks, lying down among terns, gulls and juvenile Shelducks, can be exceedingly difficult to find, particularly when hiding the head in the back feathers. 24.8. LG

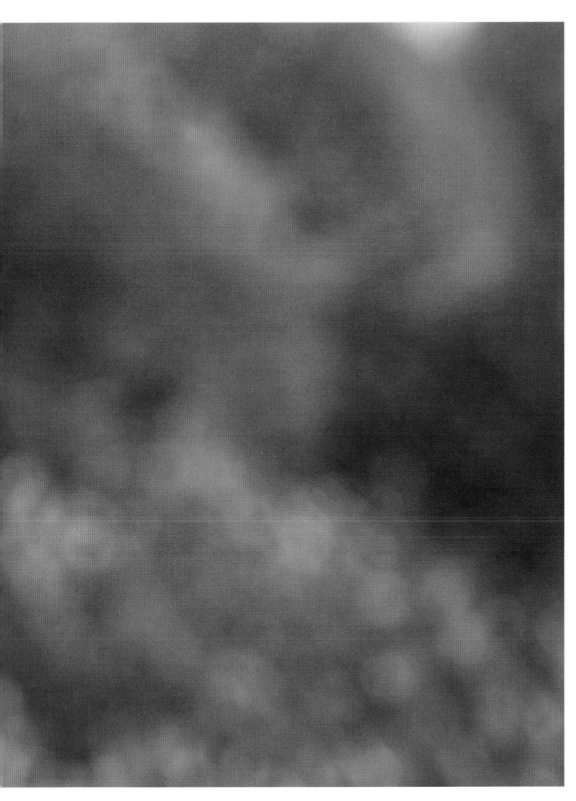

The Black-winged Stilt is the other representative of the family *Recurvirostridae* in Europe and is just as unmistakable as its relative with the upturned bill.

It does have a more conventionally shaped bill than the Pied Avocet, but makes up for this by the extremely long pink legs which have given it its name.

The Black-winged Stilt occurs throughout the world as several distinct subspecies, and the genus has another member, the Black Stilt, found only in New Zealand.

Black-winged Stilts in flight.
At the top: adult female breeding with brown scapulars.
At the front and at the bottom: adult male breeding with black scapulars and green iridescence on the upper wing.
In the middle (out of focus): first summer with diagnostic white trailing edge to the secondaries and pale tips on inner primaries. Sexed as female by brown scapulars and remiges. 22.3. Khok Kham, Thailand. HS.

BLACK-WINGED STILT
HIMANTOPUS HIMANTOPUS

Meaning of the name
'Feet on a long leash'
Greek: *himanto*, leash and *pous*, foot.

Jizz
L. 35–40 cm. Ws. *c.* 75 cm.
An unmistakeable, slim-bodied and graceful wader in black and white, with very long pink 'stilts' for legs and with a needle-thin, black bill.

In flight. Elegantly with rapid strokes of the broad wings, which have a very sharply tapering hand. The stretched-out silhouette with the extremely long legs behind the tail makes it very easy to identify the species, even from afar.

Adult male, breeding.
Note the entirely black upperside of the wing which, depending on the angle of the light, shimmers with a metallic dark green sheen. Only the primaries remain wholly black.

Plumage and identification
Adult breeding male. Head and neck may be pure white or have variable black markings; a full black hood reaching the crown, or just a black nape and back of neck. The iris is scarlet.

The body is white, the upperside of the wings black with a dark green metallic sheen.

Only the flight feathers are pure black. The undersides of the wings are black with white axillaries. The tail feathers are faintly greyish to buff-tinged.

Female. Like the male, but with a brownish-tinged back, and upper wings with no (or only faint) metallic sheen.

Adult non-breeding. As breeding plumage, but with greyish wash to crown and nape.

Juvenile. Has grey-speckled head and nape and a general greyish-brown tinge on the upperside.

The scapulars and coverts as well as tertials have beige fringes and the trailing edge of the wing is white, especially on the secondaries. Legs greyish-rose.

First winter/first summer. Differs from adults by having brown mantle, white tips on the secondaries and the innermost primaries, which are slowly worn off and shed during summer.

Voice
Does not sing but has a high-pitched two-tone call '*kyik kyik*', which may be soft or harsh and grating. A single-tone version also occurs, depending on the bird's state of mood and activity, whether courting, calling, warning or scolding.

Habitat

The preferred breeding habitat is shallow and insect-rich freshwater lakes, marshes, ponds and quiet river banks. Alternatively breeds in brackish lagoons, salty marshland and lagoons, occasionally on flooded fields, at fish farms and sewage ponds. Outside the breeding season stilts are often seen in flocks in coastal estuaries and at larger inland lakes.

Breeding biology

Can breed from one year-old. May quite often breed in solitary pairs, but most frequently in colonies of as many as 100 pairs. Spring is the time for aerial duels between boisterous males, jumping and dancing with dangling legs, in the struggle to attract mates and hold nesting territories.

The nest is well-lined, and it may be placed on the bare ground as well as on a grass tussock or even floating vegetation. The clutch of four eggs is incubated by both parents for a good 24 days. The young are able to fly after 30 days and remain with their parents for up to a couple of months.

▲ Adult female, breeding.
A brownish tinge on the upperside of the wing and brown scapulars distinguishes the female from the adult male.
Note the moulting gap in the wings, and that the inner primaries have been shed and replaced by partly grown new ones.
Also note the conspicuous white back wedge, which the species shares with curlews and some *Tringa* sandpipers. 14.7. NLJ.

▼ Adult breeding male.
The sex of the three birds portrayed here has been determined on the strength of the green metallic sheen on wings and scapulars, which are not seen in the cropped photo. 24.4. HS.

▼ Adult breeding male.
The extent of any black head markings varies from bird to bird and cannot be used for sex determination. 20.6. NLJ.

▼ Adult breeding male. 26.4. KBJ.

Migration

The majority of the European breeding population is migratory, arriving at the breeding areas in March to April. Breeding period from April to June.

Increasing numbers now winter in south-western Europe, while the majority migrate to west Africa from July to November. East European birds winter in the eastern Mediterranean and in tropical Africa.

▼ Juvenile foraging in a lake. The summer diet consists of diverse insects, aquatic animals, worms and small fish. 15.7. NLJ.

▲ Juvenile with the characteristic grey head and nape, beige fringes on the mantle, scapulars and coverts as well as white tips of secondaries and inner primaries.
    Also note the faint green metallic sheen on the greater hand and arm coverts, which indicate juvenile male. 23.7. NJL.

Distribution

The Western Palearctic population is believed to be between 51,000 and 70,000 breeding pairs. Spain is the stronghold of the species with approximately 45% of the population, followed by Russia, France, Italy and Portugal. The northern limit of the breeding distribution extends from northern Belgium and central Germany to the east. However, in recent years some pairs have successfully bred further to the north, for example in Holland, England and Denmark.

The Black-winged Stilt is found in suitable areas throughout Europe and eastwards through southern Russia and central Asia to the Pacific.

With five acknowledged subspecies scattered over all continents (except Antarctica), this species is one of the most cosmopolitan among waders.

Adult male breeding.
Lesbos, Greece. 29.4. DP.

The thick-knees form a predominantly tropical family with just one species in Europe.

The remainder of the family comprises nine species in two genera, occurring on all continents except Antarctica.

The Stone Curlew is a large, well-camouflaged wader which is difficult to observe, owing to its nocturnal ways and its earth-coloured plumage, which provides superb camouflage when it rests immobile on bare ground in the daytime.

In flight the bird is conspicuous with its pale mid-wing panel bordered by narrow black and white wingbars and the white patches in the black primaries.

The most interesting encounters with Stone Curlew are during its breeding period, at night and preferably in moonlight. At this time it is highly active with peculiar screams and courtship rituals, with the white wing patches glinting in the darkness – they no doubt have some value as a signal during courtship and the struggle for the most desirable territories. However, note that in Britain this is a scarce species and is strictly protected against any kind of disturbance near its nest. Mallorca. S'Albufera. 24.7. NLJ.

STONE CURLEW
BURHINUS OEDICNEMUS

Meaning of the name

'The ox-billed swallower'
Greek: *bous*, ox and *rhin*, bill or horn.
Oideos, to swallow.

Jizz

L. 40–44 cm. Ws. 77–85 cm. Large, earth-coloured, streaked, long-bodied, sturdy wader with a large head, staring yellow eyes, a short powerful bill and long, strong legs.

Well-camouflaged and very difficult to spot when resting in the daytime. Walks a great deal and runs very fast when pursued. Forages plover-fashion with long pauses to watch for prey, interspersed with short intervals of walking or running and swift pecking at the prey.

In flight. Uses lapwing-like flight on broad, stretched-out, slightly curved wings with characteristic pale panels on the arm and white spots on the hand. The feet do not extend beyond the tail.

Plumage and identification

The sandy or earth-coloured plumage has the same appearance throughout the year.

Adult breeding. Has a large head with big yellow eyes, marked by broad white stripes above and below the eye together with a dark moustachial stripe. The bill is short and powerful, with a yellow base and black tip. Neck and breast have narrow, dark streaks extending down on to the flanks. The remainder of the underside is white. Legs yellow and robust with a thickened ankle joint, hence the English family name 'thick-knees'.

In flight. The white tail feathers with black bars across and on the tips are seen when the bird turns and lands, as well as in courtship and defence of the territory, but are normally hidden by the long upper tail coverts.

In flight two well-marked light-coloured bars show on the arm, bordered by a dark bar together with a white speculum on the innermost secondaries. The underside of the wings is light-coloured with a dark, oval bar on the underside of the hand.

Adult male. On standing birds often only the upper and strongest marked wingbar on the median coverts is visible. In the male the white bar is flanked by two conspicuous and equally broad, blackish-brown stripes.

Adult female. Similar to male, yet more faintly coloured and with narrower lower border to the white wingbar.

Juvenile. Similar to adult, but lacks the dark borders to the white wingbar and has a less distinct white supercilium (sometimes absent).

Voice

Has a wide range of both short and long, single- or two-toned screaming calls, based on a high-pitched '*kluu-liee*'. Is very noisy and active at night, especially during the mating season.

The Stone Curlew rests during the hottest time of the day, either standing immobile or lying on its tarsi. The excellently camouflaged plumage allows the bird to merge with the surroundings. The most noticeable feature is the piercingly yellow eye. Mallorca. S'Albufera. 21.7. NLJ.

Subspecies

At present, the species is divided into five closely related subspecies.

The nominate breeds in southern England, Spain and Portugal and further east through the Balkans to the Caucasus.

Subspecies *B. o. saharae* breeds in north Africa, on the islands of the Mediterranean, in Greece, Turkey and the Middle East, while *B. o. harterti* breeds from the Volga in Russia and eastwards, *B. o. insularum* breeds on the central and eastern Canary Islands, and *B. o. distinctus* breeds on the western Canaries.

Habitat

The species is not dependent on water and the preferred habitat is open grazing fields and semi-arid, stony or sandy areas with scattered vegetation. Outside the breeding season it occurs in similar habitats.

Breeding biology

During the mating season there is boisterous activity in the breeding area, especially at night and in the morning. Shrieking calls and duets form part of the acoustic scene during the dark hours. In daylight the behaviour includes running and screaming threats and ritualised bows and turns. The nest is built on the bare ground with little or no lining, often decorated with pebbles or shells along its edges. The sole clutch of two eggs is incubated by both parents for 25 days or more. After hatching, the chicks follow the parent birds and are able to fly and live independently after 40 days. The diet is wide-ranging, consisting of small animals, from snails and spiders to insects, amphibians, lizards and mice.

Migration

Birds from the northern and eastern European populations move to winter quarters in southern Europe, the Middle East and Africa, while the southern European populations are primarily sedentary.

▲ Shrill screams, threatening behaviour, standing erect and bowing are constant activities during the breeding season, as part of territorial defence and courtship. The bird to the right is moulting.
The bird to the left is an adult breeding male, identifiable as such by the distinct dark borders to both sides of the white wingbar across the arm. Mallorca. S'Albufera. 23.7. NLJ.

▼ Juvenile with fluffed-up feathers. Age is determined by the faintly marked supercilium and the fresh plumage with broad, light beige edges. Mallorca S'Albufera. 23.7. NLJ.

Distribution

The European population has been estimated at between 50,000 and 82,000 breeding pairs. The northernmost outpost is a small population in southern England. Patchily distributed in suitable habitat over the remainder of western Europe.

The largest and most stable populations are in Spain, Portugal and France. The range extends from the west through eastern Europe, southern Russia and Kazakhstan to north-western China.

The Glareolidae family comprises the lapwing-like coursers (represented in Europe only by the Cream-coloured Courser) and the tern-like pratincoles.

The nine species of coursers display a plover-like behaviour and pigeon-like flight. They do not breed in continental Europe and only one species is recorded, as a rare vagrant.

The coursers have a predominantly African distribution, with seven species on the African continent. One of these is also found further east, through the Middle East and Arabia into India, and there are a further two species in southern Asia.

The elegant long-winged pratincoles number eight species, of which five are primarily African, two Asian and one Australian.

Two species breed in Europe and a third is a rare vagrant from Asia.

Cream-coloured Courser, adult breeding.
   Three adults in their breeding area in typical habitat of semi-desert or true desert.
   Immobile birds virtually merge with the beige desert scene and may thus be very difficult to spot.
   They mainly move about by rapid scurrying. They feed mostly on insects and other small creatures, which are spotted and snatched in plover-like fashion. Nitzana, Israel. 9.4. LK.

CREAM-COLOURED COURSER
CURSORIUS CURSOR

Meaning of the name
'Runner'
From *cursorius*, courier, *currere*, to run.
Cursor, synonym of *cursorius*.

Jizz
L. 19–22 cm. Ws. 51–57 cm.
A medium-sized, atypical wader with large black eyes, a short curved bill, medium-length neck and long legs. The posture is erect, almost vertical in alert birds and the behaviour appears to be shy and evasive with keen watchfulness and brief, fast runs.

In flight. Pigeon-like, with deep chest, rapidly swinging strokes and intermittent short glides.

▲ Adult breeding. The Cream-coloured Courser is unmistakeable when standing as well as in flight, when the beige body and contrasting black hand and black underwings stand out at a great distance. Note the thick soles under the toes, presumably an adaptation to cope with the blazing hot desert sand. 26.4. Oman. HJE.

Plumage and identification

Adult. Male and female have a similar appearance throughout the year.

The plumage is unmistakable with its cinnamon-coloured forehead and greyish-blue hind-crown, which is framed by a broad white supercilium and a black eye-stripe which converge on the nape in a striking black-and-white triangle. The bill is curved, pointed and bicoloured, black with grey base.

Birds on the ground show uniform sandy-buff upperparts, beige-tinged white underside and long whitish legs. In flight and other plumages. Refer to photo captions.

Subspecies

Besides the nominate *C. c. cursor*, there are two further subspecies. *C. c. excul* on Cape Verde and *C. c. bogolubovi* from southern Turkey and eastwards through Iran and Afghanistan into India.

◀◀ Non-breeding.
The bill, the head markings and the plover-like alert behaviour make the Cream-coloured Courser unmistakable both among desert birds and waders. 24. 12. Oman. HJF.

◀ Juvenile.
Easily distinguishable from adults by its whitish-buff appearance with faint dark vermiculation on crown and upperside. The vermiculation is soon worn off, and in late autumn the juvenile moults into adult-like plumage.
During a brief period, first winter birds can be recognised by buff-edged primary tips, which are here seen extending beyond the tertials. 30. 5. Golan, Israel. LK.

▶ Adult non-breeding, stretching its wing. Seen from behind, the characteristic nape pattern is remarkably vivid, and on the stretched-out wing so are the bicoloured secondaries in black and buff with white tips, which are so striking in flight. 23.12. Oman. HJL.

Habitat

Breeds from February to May. Ideal habitat is hot, dry desert or semi-desert, steppe and saline plains.

Migration

Birds of the nominate form breeding in northern Africa undertake intra-African migrations across the Sahara, to winter quarters in the Sahel zone. Lost birds have turned up in Europe, including Scandinavia, as vagrants, mostly in autumn.

Distribution

The European population consists of between 200 to 2,300 breeding pairs on the Canary Islands (Lanzarote and Fuerteventura). The nominate breeds from the Canaries across north Africa to the Arabian Peninsula.

Voice

The flight-call is a repeated, subdued and coarse *'quett'*.

Collared Pratincoles on spring migration. In favourable light it is easy to make out the five significant flight features: brick-red underwing; long forked tail; white trailing edge of the arm; a distinct contrast between the black hand and the greyish-brown upperside; and the white-shafted outer primary.
La Janda, Andalusia, Spain. 27. 4. HS.

In Europe, the pratincoles are represented by two species: the Collared Pratincole and the Black-winged Pratincole.

Both species are slender, graceful and short-billed and, in their behaviour and appearance, very different from other waders, particularly in flight when their similarity with both terns and swallows is striking.

These elegant birds are heat-loving and found in southern and eastern Europe, in insect-rich meadows and steppes close to wet areas.

Their food consists primarily of large insects such as grasshoppers, beetles and dragonflies, which are mostly caught in the air, most often in the morning and at dusk when the typical silhouette of the pratincoles is clearly set off against the sky.

Unfortunately, the pratincoles are under pressure because of agricultural intensification, particularly draining of wetlands and use of pesticides, both in the breeding areas and in the winter quarters in Africa.

A further six species occur in Asia, Australia, Africa and Madagascar.

COLLARED PRATINCOLE
GLAREOLA PRATINCOLA

Meaning of the name
'The settler at the meadow'
From *glareola*, gravel. Presumably reference to the gravelled breeding areas.

Prati, meadow, *incola*, settler.

Originally listed among the swallows by Linneus, later transferred to the present order.

Jizz
L. *c*. 22–25 cm. Ws *c*. 65 cm.

Resting birds resemble brownish-buff, short-billed terns on long, dark legs.

Often shuffles around but is capable both of running and jumping high in pursuit of small animals and insects.

In flight. Has deeply forked tail. In regular flight, such as during migration, which often occurs in large flocks, the tern-like jizz is again evident. However, during foraging for insects the flight becomes livelier and more acrobatic, suggestive of a large swallow on very long, pointed wings.

Plumage and identification
Adult breeding. Bill black and curved. A red base on the inner half of both upper and lower mandible produces marked 'painted' corners of the mouth. The creamy brown chin is separated from the greyish-buff breast by a narrow, black chin-band. The upperside is fawn with a white rump. White belly and greyish-black legs. The tail, with its greatly lengthened outer feathers, protrudes a little beyond the folded wings, thus distinguishing the species from both Black-winged and Oriental Pratincoles.

However, note that as a result of wear and tear the tips of the tail feathers are sometimes the same length or shorter than the wing tips.

Male and female. The only visible difference between the sexes is the lore which in the male is black and in the female brownish.

In flight. What sets the species apart from the Black-winged Pratincole in flight are the red armpits and arm coverts of the underwing, together with the contrast of the upper wing between the black hand and the greyish-brown colour of the arm and upperside.

▲ Adult breeding. The pale trailing edge to the secondaries can be seen only from beneath and during the breeding season it diminishes due to wear. These pale shaft-streaks, particularly on the outermost primaries, are characteristic of the species. The Black-winged Pratincole has dark shaft-streaks, and the Oriental Pratincole toned-down buff ones. 2.6. KF.

▼ Adult male, breeding. The difference in shape and size of the nostrils in the three species dealt with here is allegedly a useful field-mark. Close studies of the photo material for this book do not support this theory, suggesting that differences in shape are down to individual variation. 27.4. HS.

The Black-winged Pratincole, on the other hand, has a uniform dark upperside lacking contrast. Besides, the outermost primary has a marked white shaft, and in fresh plumage a broad light-coloured trailing edge to the secondaries.

Adult non-breeding. Develops a faintly streaked head and throat as well as thin white fringes on mantle, scapulars and coverts. The red colour on the bill fades into orange, and the distinct black throat-surround is somewhat blurred.

Juvenile. Light-coloured on head, throat and breast with dark streaks. Mantle, scapulars and coverts are greyish-brown with black-tipped, buff fringes.

Voice

The typical call is a tern-like shrill and vibrating '*kitt*' or '*kitt-e-litt*', also used in a brief trilling version as an alarm call. In spring the call is developed into a song, consisting of rhythmical multi-syllable sequences, also heard from migrating flocks.

Habitat

This species prefers open spaces such as meadows, arable land and grass steppes with ample insect populations, often near wetlands. Its strongholds are in Spain, Portugal, Greece, Romania, Russia and Azerbaijan. Breeding season is from May to August.

Breeding biology

Breeds in flat open country in meadows and grass steppes both near brackish coasts, salt marshes and inland lakes, in groups numbering from between a few pairs to as many as 100.

The nest is scraped out or built in a depression in the ground. The 2–4 eggs are hatched in *c.* 18 days, both parents incubate.

The chicks stay in the nest for the first couple of days, and are fed by the parents for up to a week with both fresh and regurgitated food in the form of insects, which are either snatched on the ground or

▲ Adult breeding. The contrast between the dark hand and the paler, greyish-brown back and upper wing, as well as the white trailing edge to the secondaries, which is most conspicuous in fresh plumage, distinguish the species from both Black-winged Pratincole and the much rarer Oriental Pratincole. La Janda, Andalusia. Spain. 30.4. HS.

▲ Adult breeding. In glaring light as here, the brick-red feathers of the armpit and underwing coverts can be very hard to tell from the rest of the dark underside of the wing. Shortened tail feathers and loss of the pale trailing edge to the arm (through wear) may complicate the identification of Collared and Black-winged Pratincole. 6.6. NLJ.

caught in the air, preferably above water. Foraging takes place in the early morning or late evening.

The young are able to fly after 25–30 days.

Migration

European birds migrate to the winter quarters in Africa south of the Sahara during August–October and return during April–May.

Distribution

There are between 6,300 to 11,900 European breeding pairs. The nominate *G. p. pratincola* breeds patchily in suitable habitat in western and eastern Europe and through Kazakhstan to north-western China. The largest European populations are found in Spain and Russia. It also breeds in north Africa, Turkey, parts of the Middle East and Pakistan. Similar subspecies are found south of the Sahara.

▼ Juvenile. Distinguished from the closely related juvenile Black-winged Pratincole by the red corners of its mouth and an altogether paler and less contrasting appearance. 8.9. Oman. HJE.

BLACK-WINGED PRATINCOLE
GLAREOLA NORDMANNI

Meaning of the name
'Nordmann's Pratincole'
From *glareola*, gravel. Presumably referring to gravelly breeding areas.

Nordmanni, from the Finnish zoologist, Alexander von Nordmann. (1803–1866)

Jizz
L. *c.* 23–26 cm. Ws. *c.* 65 cm.
A slender, brownish-beige, tern-like and short-billed wader on medium-long legs.
Picks up insects and small animals on the ground in brief runs and jumps into the air for low-flying food items.

In flight. Has a forked tail and tern-like flight on long, pointed wings, with swallow-like plunges and turns during foraging, which is often carried out by flocks.

Plumage and identification
Standing birds look deceptively like Collared Pratincoles, but are a tiny bit darker and in fresh plumage distinguishable from it by the elongated tail feathers, which are shorter and never project beyond the folded wing-tips.

In flight the entire underwing is blackish-brown, and the upperside is dark without any significant contrast between hand and arm. The species has no pale trailing edge to the arm.

Adult breeding. Male larger than female with broader lore-stripe which breaks the white eye-ring and extends to a bit over the eye.

Adult non-breeding. Attains faintly streaked head and throat along with narrow white fringes on the mantle, scapulars and coverts. The red colour on the bill fades to orange-red, the throat fades and the clear-cut framing becomes more blurred.

Juvenile. Almost indistinguishable from the juvenile Collared Pratincole. Has an entirely black bill and is darker and more boldly patterned.

First winter/first summer. Distinguished from adult by narrow edges to the inner secondaries.

▲ Adult breeding.
In flight displays black armpit feathers and totally dark underwings. The upper wing is dark with a faint contrast between the dark brown back and arm and the nearly black hand. Lacks the white trailing edge of the arm that is seen in Collared Pratincole. South Africa. 1.3. WRT.

▼ Adult breeding.
The bird is worn and beginning to attain the adult non-breeding plumage, with a faded chin area and unevenly demarcated black throat-collar.

Note the elongated but actually relatively short tail feathers, which fall some way short of reaching the tips of the folded primaries. 15.7. Kazakhstan. SD.

Voice
Similar to that of Collared Pratincole, but the call is a shorter, harsher and more crackling '*kee-tick*'

Habitat
Found on damp meadows close to water and on drier shores of lakes and inlets with plenty of insects.

Breeding biology
Preferred habitats are the banks of lakes and rivers, farmland and salty meadows and grass steppes close to water.

A colonial breeder, ranging from a few pairs to nearly a hundred. The nest is placed in a simple scraped-out hollow and sparsely lined. The eggs, up to four, are incubated by both sexes for presumably 18 days, as with the Collared Pratincole.

The chicks are independent at about 30 days old. The food consists of insects and other small creatures, primarily caught in the early morning or late in the evening.

Migration
Arrival at the breeding grounds takes place in May–June, and after the breeding season the birds gather in large flocks for the migration to southern Africa. Migrates in August–October over a broad front, and at great height.

Distribution
The total European population of 6,000 to 7,000 pairs is almost entirely confined to Russia. The range extends from Ukraine through Russia and eastern Kazakhstan eastwards to Mongolia.

▲ Non-breeding, with a grasshopper.
In their winter plumage standing birds are very difficult to distinguish from Collared Pratincole.
Both species have streaked heads, faded dark corners of mouths and thin white edges on the newly grown feathers. The Black-winged Pratincole can be recognised by the shorter elongated tail feathers, which never reach the folded wing tips. 10.10. Israel. LK.

▼ Juvenile. Compared to the Collared Pratincole, the Black-winged Pratincole has a more contrasting plumage and an all-black bill. Juvenile Collared Pratincole has dull reddish corners to the mouth.
Post juvenile. Depending on the time of hatching, Black-winged Pratincole may have reddish mouth-corners already in late summer. 15.7. Kazakhstan. SD.

ORIENTAL PRATINCOLE
GLAREOLA MALDIVARUM

Meaning of the name
'The pratincole from the Maldives'
From *glareola*, gravel.
Maldivarum, Maldives.
Named by the English doctor and ornithologist, John Latham (1740–1837), from a specimen taken on a ship in the Indian Ocean on the degree of latitude with the Maldives.

Jizz
L. 23–24 cm. Ws. *c.* 57 cm.
Like the other pratincoles, a short-billed tern-like wader with medium-long legs. Most closely resembles a slightly shorter-bodied and slimmer Collared Pratincole, with less elongated tail feathers.

In flight. Long-winged and tern-like with buoyant flight. Has forked tail with significantly shorter outer tail feathers than both Collared and Black-winged Pratincoles.

Similar species
Very similar to a short-winged and short-tailed Collared Pratincole.

Plumage and identification
Adult breeding. On resting birds the outer (longest) tail feathers fall significantly short of the tips of the folded wings.

The transitional area between the greyish-brown breast and the pale belly is of a similar orange-buff colour as the throat bib, and not nearly as contrasting than that of the Collared Pratincole.

In flight. In flight the upperside is less contrasting than on Collared Pratincole and differs from it in lacking the white trailing edge to the secondaries. The underwing has brick-red arm coverts and armpit feathers. No visible difference between the sexes but the male has a more deeply forked tail.

Adult non-breeding. As with the other two pratincoles, this species attains a faintly streaked head and throat as well as narrow pale edges on mantle, scapulars and coverts. The red colour of the bill fades to orange-red, the throat becomes paler and its black and white border becomes more diffuse.

▲ Both birds in flight are adult breeding.
In flight resembles a short-tailed Collared Pratincol, lacking the long tail-streamers. The fork of the tail is less deep on the left bird than normal, due to worn-off feather tips.

Note that the species has almost no contrast on the upperwing, unlike Black-winged Pratincole, and also lacks the prominent white trailing edge to the secondaries so typical of the Collared Pratincole.

Furthermore, the shaft streak on the outermost primary is faintly cinnamon-coloured and not white as on Collared, while the brick-red part of the underwing is of the same appearance to Collared. 24.4. Petchaburi, Thailand. HS.

◄ Adult breeding.
Resembles a short-tailed and slender version of Collared Pratincole. Note the short tail-streamers which are just visible below the tips of the primaries. 24.4. Petchaburi, Thailand. HS.

Juvenile. Has contrasting, scaly appearance on the coverts and the feathers of the upperside caused by buffish fringes and pale tips of the primaries. Resembles juvenile Collared and Black-winged Pratincoles, but has very short tail feathers.

Voice

Short, vibrating and rolling '*preek*' or '*prett-prett*'. Slightly softer and less shrill than that of Collared Pratincole.

Habitat, migration and distribution

Very rare vagrant from south-eastern Asia to Europe; recorded in Sweden, Denmark, England, the Netherlands and Cyprus. Asian birds migrate south to the winter quarters in southern Asia, through Indonesia and New Guinea to Australia.

Northern Lapwing and Common Ringed Plover, adult breeding. The photo illustrates differences in size and variation in plumage within the family of plovers, which contains the second highest number of species of all wader families. 11.4. HS.

The banded plovers, the Pluvialis plovers and the lapwings together constitute the family Charadriidae. This is the second largest wader family after the sandpipers and their allies, Scolopacidae.

The plovers occur on all continents except Antarctica. There are 12 genera with 71 species as well as numerous subspecies.

Ten species breed in Europe and an additional seven are recorded as rare vagrants.

The common denominators for the small to medium-large plovers are their short, stout bills, large head with large eyes and a characteristic feeding technique consisting of three typical stages: watch, run and snatch!

The prey is found by sight, reached by a short run, then caught and swallowed.

Additionally, a 'pumping' or 'paddling' method is used on damp, grassy areas, whereby the bird stamps with one foot, creating vibrations in the upper surface of the soil. Theory has it that the vibrations mimic the effect of heavy rain, thus encouraging the earthworms to move upwards towards the surface in order to avoid drowning. In shallow water, one foot is rapidly fluttered to expose food items in the topmost layer of mud.

117

Common Ringed Plover. Adult male breeding. From a vantage-point in the territory the male anxiously warns the chicks against two-legged intruders. 7.7. LG.

The banded plover genus, *Charadrius*, numbers 33 species and is the largest of the wader genera.

In Europe only four species breed, and also the Dotterel, *Eudromias morinellus*, which is the only species in its genus but is related to the banded plovers.

Additionally, four species are rare vagrants from North America and Asia.

The banded plovers are the smallest of the plovers, slightly smaller or the same size as Dunlin.

As a group they are easy to tell from other waders because of their plain greyish-brown upperparts, white belly and, in particular, by the characteristic head markings and the breast-band which is either black or reddish.

The banded plovers nest on gravelly, stony and sandy ground by water, as well as on saline steppes and semi-desert.

Outside the breeding season they occur by lakes, rivers and coasts.

COMMON RINGED PLOVER
CHARADRIUS HIATICULA

Meaning of the name
'The creek-dwelling plover'
Charadrius, name for plovers, derived from Greek, kharadra, creek, with reference to bird's habitat along gravelly streams in creeks.

Hiatus, creek or opening, and *cola/colette*, inhabitant.

Jizz
L. 18–20 cm. Ws. 35–41 cm.
The largest, most robust and boldly marked of the three European banded plovers.

As in all banded plovers, the body is elongated and deep-chested with a large, round head and large eyes. When disturbed it often crouches and runs rapidly close to the ground, then pauses before running again, before it takes flight.

In flight. Flies swiftly and straight, often with turns. In the breeding season the male overflies the territory with stiff 'rowing' wing-beats, in circles or figures of eight.

Similar species
Semipalmated Plover. See page 124.

Plumage and identification
Adult breeding, male. Has large dark eyes with very narrow, almost invisible yellow orbital ring. The forehead is white, the bill orange with its outer third black. Crown, nape and upperside brownish-grey. Facial mask and breast-band black. Rest of breast and belly white. Legs are orange.

In flight. Has broad, white wingbar, resembling staircase steps on the primaries. The tail feathers fade from greyish-brown into black with white tips and white outer tail feathers.

Female. As male, but more greyish with brownish-black head markings and breast-band.

Adult non-breeding. Loses the black frontal bar and generally appears more greyish-brown. The breast-band is reduced and may appear broken with dark spots on the centre of the breast. The

▲ Adult breeding, male.
Common Ringed Plover is similar in size to the medium sized *Calidris* species and likewise has a white wing-bar, but stands out in mixed flocks of waders by colour and shape of its bill as well as by the black head markings and breast-band. 17.8. NJL.

▼ Adult breeding, male.
Differs from the female by the more saturated black markings and on the more brightly coloured bill and legs. 7.7. LG.

▲ Juvenile.
May resemble juvenile Little Ringed Plover which has lemon-yellow orbital ring and a very narrow white wing band, most pronounced on the arm. 17.8. NLJ.

bill is dark with faint reddish-orange markings at the base of the lower mandible. Legs dull orange-yellow. In early winter the black markings again become more prominent.

Juvenile. Finely pale-fringed on mantle, scapulars and coverts and with pale edges to the tertials. Lacks black frontal bar but has black bill. The breast-band is broken and the legs are brownish.

Subspecies

The nominate form, *C. h. hiaticula*, breeds from the British Isles and southern Scandinavia south to north-western France, along the Baltic coast and locally in Poland.

Subspecies *C. h. psammodromus* breeds from north-eastern Canada through Greenand to Svalbard, Iceland and the Faroe Islands, and *C. h. tundrae* breeds from northern Scandinavia along the Russian and Siberian coasts across the Bering Strait to north-western USA.

▲ Adult breeding, male. *C. h. tundrae*.
Notice more narrow white frontal bar and finer bill compared to the adult nominate form to the left on the opposite page.
Both psammodromus and tundrae are smaller than the nominate form but difficult to distinguish unless the subspecies are seen side by side on migration, where the difference in size is apparent.
Oppdal, Norway. 11.6. HS.

▼ Adult breeding, female.
The vast majority of adult females are pale greyish-brown with blackish-brown admixed in the black markings, and have more subdued coloration of legs and bill. Thus the female blends perfectly with the surroundings when she is incubating among stones, mussel shells and remains of seaweed. 7.7. LG.

Voice

The flight and alarm call is a soft, anxious 'poo-ee'. The song is creaky, monotonous and rapidly repeated 't'weea-t'weea', which is most often delivered from a characteristic 'rowing' song flight circling over the territory, both over water and land.

Habitat

The species can be seen from late winter and early spring, singly or in small groups on muddy, sandy and stony coastal stretches. Breeds on gravelly and stony beaches, on coastal meadows and more rarely on sandy fields and by gravel pits and inland lakes.

Breeding biology

Depending on the degree of latitude, breeding is from April to July. Breeds when one year old and stays with the same partner for one or several seasons.

Pairs breed singly or in loose colonies with ample distance between the nests.

The nest is placed in a scrape on the ground and is scantily lined with smaller pebbles, fragments of mussel shells or plant material.

Southern populations lay 1–3 clutches, including re-laying after predation or flooding. Northern populations have only time for one clutch.

The eggs, usually four, are incubated by both sexes for c. 25 days. Chicks follow their parents after hatching, searching for insects, worms and small crustaceans which are snatched after a short sprint.

The young fledge after 24 days and are then almost independent.

The oldest ringed bird attained the age of at least 20 years.

▲ Adult breeding.
A singing male with fluffed out throat lands on the breeding ground. The territory is claimed by means of a song flight delivered in a typical, stiff-winged, 'rowing' flight in circuits often low over the ground. 4.6. NLJ.

▼ Adult breeding male.
Only about 61% of the newly hatched chicks survive until fledging age because of predation, mainly by gulls and corvids. 4.6. HS.

Migration

The spring migration to and through Europe takes place along the coasts as well as over a broad front over land. The migration of the nominate, *hiaticula*, which breeds in northern and western Europe, culminates in mid-April. Return migration towards the winter quarters begins from the end of June, along the European coasts of the Atlantic, primarily in southern England, Ireland, France and Spain.

The north-eastern and eastern populations of the subspecies *tundrae* migrate towards the breeding areas in northern Scandinavia and northern Russia during May–June, to return again by August. Western populations winter along the coasts of the Mediterranean and further south to South Africa. The eastern populations migrate from the Caspian Sea to south-western Asia and south to South Africa.

Birds of the subspecies *psammodromus* that breed in north-eastern Canada, Greenland and Iceland migrate across the Atlantic, and down along the coasts of western Europe towards their winter quarters in south-western Europe and Africa.

Distribution

The European population amounts to between 110,000 and 253,000 breeding pairs. Breeds in Europe from the British islands and north-eastwards through northern Europe and the mountains of Scandinavia and northern Russia.

Distributed further east through the entire Siberia to the Pacific.

▲ Juvenile.
The juvenile plumage lasts through autumn until the beginning of winter when the pale fringes are worn off and the bird resembles a winter adult. First winter birds are thereafter only distinguishable from adult non-breeding by retained pale fringes on the inner median coverts and on worn, pointed tertials.
   Note that the similar juvenile Little Ringed Plover, *Charadius dubius*, has yellow orbital ring and buff fringes. 3.9. JG.

▶ Non-breeding.
Note dark bill with faint reddish on the innermost part of the lower mandible, lack of black frontal bar and incomplete black collar. 3.1. Spain. DP.

SEMIPALMATED PLOVER
CHARADRIUS SEMIPALMATUS

Meaning of the name

'The plover with palm-of-the-hand-like toes'
Charadrius, name for plovers, derived from the
Greek *kharadra*, creek, hinting at the bird's habitat
along gravelly streams in creeks.

Semi, half, *palmatus*, palm of the hand. Relating
to the partially webbed toes.

Jizz

L. 17–19 cm. Ws. 43–52 cm.

The North American counterpart of the Common
Ringed Plover is a trifle smaller, has a rounder head,
a shorter bill and a more compact appearance.

Plumage and identicaition

Extremely similar to Common Ringed Plover and
can only be safely identified by using a combination of the head markings, the shape of the bill, partially webbed toes and the voice.

Size of the body, details in the plumage of the
upperside and the extension of black on the bill do
not make the basis of a definite identification on
their own.

Individuals from the Siberian population of
Common Ringed Plover, *C. h. tundrae*, which are
seen on migration in western Europe, appear small
and dark, closely resembling Semipalmated Plover.

Furthermore, some individuals of Common
Ringed Plover (of both subspecies) have a vaguely
defined yellow orbital ring and hints of webbing
between the middle and the outer toe.

Head markings, adult breeding, male. Has
distinct yellow orbital ring. The bill is slightly shorter and more delicate, and the white patch over and
behind the eye is ill-defined and mostly does not
have the appearance of a definite supercilium as in
Common Ringed Plover.

▲ Semipalmated Plover. Adult breeding, male.
The yellow orbital ring, the vague whitish area over
the hind part of the eye, narrow white frontal bar,
narrow black collar and a short, dainty bill are typical
identification marks of the species. However, the head
markings are quite variable as to the extent of the black
and the white frontal bars, or whether the black on the
forehead is sharply demarcated from the crown or not.
Likewise the supercilium may be quite well-defined and
the bill is sometimes all black. Canada. 6.6. DP.

▲ Common Ringed Plover *C. hiaticula*. Adult breeding,
male.
The oval head shape and the sturdy bill of the Common
Ringed Plover, as seen here, are important criteria when
distinguishing between the two species.

The boldly marked supercilium seems to be a
constant feature of Common Ringed Plover, while the
other markings are more variable. Additionally both
sexes may have a very narrow yellow orbital ring in the
breeding season. 7.6. JL.

▼ Semipalmated Plover. Adult breeding, male. (Same bird as to the left above)
Note that this male has very little sign of webs between the inner and the middle toe on the left foot. It is thus
crucial to obtain good views of the entire foot. Canada. 6.6. DP.

▲ Semipalmated Plover has distinct webbing between the middle toe and both the outer and the inner toe.
Common Ringed Plover may have weakly developed webs between the middle and the outer toe. Florida. 30.12. DP.

▲ Adult breeding, female.
As in several species of banded plovers, the female shows less contrast and is paler than the male. Canada. 8.6. DP.

▲ The juvenile closely resembles Common Ringed Plover, but differs in attaining a hint of a yellow orbital ring even at this tender age. Stone Harbour, New Jersey. 5.10. HS.

▼ Adult non-breeding.
Has yellow orbital ring throughout the year, most distinct in the breeding season. Note the webbing at the outer toe.
Stone Harbour, New Jersey. 16.10. HS.

Habitat, migration and distribution

The species breeds across the North American continent in the sub-Arctic and Arctic zones. It primarily winters along the shores of southern North America as well as along the coasts of South America. Occurs as a rare vagrant along the European shores of the Atlantic; has been recorded in Ireland, on the west coast of Norway, in Britain, the Netherlands and Spain. Most records are from the Azores with almost 200 observations. This species is presumably often overlooked due to its similarity to Common Ringed Plover, and because of the severe challenge of a safe identification which demands exceptionally good viewing conditions.

Voice

The call, which is the safest identification feature, is a weaker, longer and more shrill '*plee-wee*', opposed to the softer, more plaintive '*poo-ee*' of Common Ringed Plover.

The intonation may vary from neutral between feeding birds to a more direct, bolder and shorter call, very similar to that of Spotted Redshank.

Here both call and flight call are heard.

LITTLE RINGED PLOVER
CHARADRIUS DUBIUS

Meaning of the name

'The dubious plover'

Charadrius, name for plovers, derived from the Greek *kharadra*, creek, relating to the bird's habitat of gravelly streams and shores.

Dubius, doubtful, relating to its doubtful placing in the taxonomy as the describer thought that the bird was possibly a geographical variation of the Common Ringed Plover.

Jizz

L. 14–17 cm. Ws. 42–45 cm.

Smaller and shorter-legged than Common Ringed Plover and with more slender and elongated body due to the long wings. Generally leads a quiet existence and is therefore often overlooked. This species often steals quietly away when disturbed.

In flight. Resembles Common Ringed Plover, but is more slender with more pointed wings, longer tail and more erratic *Tringa*-like flight.

▲ Adult breeding. Male in song flight with puffed-up throat. Note that when in a fresh plumage the species has two very thin wing-bars formed by the pale tips of the greater coverts on arm and hand, and of the secondaries. 18.4. LG.

▲ Juvenile has distinctly scaly appearance because of pale buff fringes of the feathers on the upperside. Lacks black frontal bar and has well-defined orbital ring, here bright whitish-yellow. Note the narrow wing-bar which is stronger than on adults. 10.8. NLJ.

Plumage and identification

Adult breeding. Has pale greyish-brown upperside. The head markings set the species apart from the other species of banded plovers by the black frontal bar being separated from the crown by a thin white line (although this feature may be absent in some individuals). Furthermore, it sports a bold, lemon-yellow orbital ring and a black bill with just a hint of reddish-yellow on the innermost part of the lower mandible. The white underside is separated from the throat and neck by a black collar. The legs are reddish-brown to greenish-yellow.

In flight. The upperwing appears uniform, lacking the prominent white wing-bar seen on Kentish and Common Ringed Plovers.

The tail has black markings and white outer tail feathers.

Male and female. During the breeding season the black markings are suffused with brownish on the females.

Non-breeding. In early winter the head is greyish, lacking the black frontal bar, and the orbital ring is pale and inconspicuous. From mid winter the black markings increase and the orbital ring regains its striking yellow colour.

▶ Adult breeding. Male with bold black markings. The habitat is typical for the species both during migration and when breeding, most often found at inland localities at lakes and gravel pits. 10.8. NLJ.

Juvenile. Quite similar to juvenile Common Ringed Plover, but has brownish upperside with buff fringes, not greyish with whitish fringes. The orbital ring is conspicuous, either very pale or yellow as in adults, but the white frontal bar is small and white.

Subspecies

Divided into three acknowledged subspecies.

Only *C. d. curonicus* (*curonicus*, Courland/Latvia) occurs in the Western Palearctic. The other two subspecies occur in south-east Asia.

The nominate, *C. d. dubius*, is found in the Phillipines and in New Guinea.

Voice

The call is a hard and short '*piu*', markedly different from the softer and anxious '*poo-ee*' of Common Ringed Plover.

The song is a shrill, tern-like '*reeee-oo*' repeated in series from slow-winged, rowing flight in circuits forth and back over the territory.

Habitat

The preferred habitat is inland localities with fresh water such as gravel pits, quarries and lakes with muddy and often dried up edges. Also found on brackish lagoons, estuaries and quiet rivers with small islands with stony or gravelly banks. Not as commonly seen on sea-shores as its two European relatives, Common Ringed Plover and Kentish Plover.

Breeding biology

Breeds when one or two years old and pairs up for one or several seasons. Breeds singly or in loose colonies, with ample distance between the nests.

The nest is a scrape on the ground, lined sparsely with smaller pebbles, mussel shells or plant material, preferably near water.

▲ Adult breeding. Little Ringed Plover and male Common Ringed Plover. Even though the image is slightly hazy the differences of size as well as the diagnostic features of the two species are clearly seen. The Common Ringed Plover has accidentally entered the territory of its smaller relative and has raised its tail in order to deter the two threatening 'land-owners'. 28.6. JL.

▼ Adult breeding, female. Sexed on the pale feathers of the ear-coverts and the brown touch to the black collar. Note the thin white line between the frontal bar and the crown, and the diagnostic white outer tail feathers with inner greyish-black markings, not seen in the two other species. Little Ringed Plover often behaves aggressively in its territory, here shooing away a Wood Sandpiper, probably in a dispute over food. 3.5. HS.

The Little Ringed Plover is highly territorial and aggressive towards other species, but sometimes chooses to nest adjacent to larger species, relying on their vigilance to keep predators away.

The usual clutch-size is four and the eggs are incubated by both sexes for *c.* 25 days. Both parent birds raise the young, but the female is the first to leave the family.

The young are able to fly at the age of *c.* 27 days and become independent after an additional one to three weeks.

The oldest ringed individual lived for at least 13 years.

Migration

The return migration to Europe from the winter quarters in the tropics takes place from mid-March to May. The females leave the breeding areas from June, later followed by the males and the juveniles. A minority may winter in the breeding areas, but the greater part of the western and eastern population migrates to the south-west and the south-east respectively in September. The former winters in the Sahel south of the Sahara, the latter around the Mediterranean, in Egypt, the Middle East and in east Africa.

Distribution

This species is the most numerous banded plover in Europe. The population is estimated at between 133,000 and 157,000 pairs. The European subspecies of Little Ringed Plover, *C. d. curonicus*, occurs from southern England through the whole of Europe including southern Scandinavia and Finland and further eastwards through Asia to the Pacific.

▲ Juvenile. The feathers of the upperside are greyish-brown with a narrow, dark subterminal band and beige fringes. White-tipped greater coverts form a narrow but distinct white wing-bar in flight.

The marked yellow orbital ring, which can be very pale, distinguishes it from all the other juvenile banded plovers in Europe. 5.7. JL.

▼ Adult non-breeding. Note the renewed dark primaries, particularly P9, which is emerging, and the faded, greyish-brown outermost P10 which has not yet been shed.

The four old inner secondaries, which are faded, are the last to be moulted.

Note that the newly emerged greater and median coverts are of different lengths because some of them are not yet fully grown. Their pale edges are worn off during the spring, so that the wing by then shows no or only a faint wing-bar. Oman. 18.11. HJE.

KENTISH PLOVER
CHARADRIUS ALEXANDRINUS

Meaning of the name
'The plover from Alexandria'
Charadrius, derived from the Greek *kharadra*, creek, relating to the plover habitat of gravelly streams and creeks.

Alexandrinus, the species name, refers to the coastal city Alexandria in Egypt, one of the the places where the species lives.

Jizz
L. 15–17.5 cm. Ws. 42– 45 cm.
A small, very pale banded plover with a broken breast-band. The bill is black and slight, the body squat and the legs are long and greyish-black, capable of fast sprinting on sandy areas.

In flight. The distinct white wing-bar, the pale greyish upperside and the dark-centered, white-edged tail are the specific features seen.

▲ Adult breeding. Male stretches its wing exposing the prominent white wing-bar, which the species shares with the Common Ringed Plover.
Note also the flat crown. The other species have a more rounded head. 26.5. LG.

▼ Adult breeding. Male in fresh breeding plumage which is attained during early winter. The intensity of the ochrous-coloured markings on the crown and nape is individually variable as can be seen on the two photos on this page. 31.12. Spain. DP.

Plumage and identification
Adult breeding, male. The rusty-ochre markings on the crown and nape vary from dull greyish-buff to a much richer hue. The bill is slender and black as are the frontal bar, the lore and the eye-stripe. Thin white eye ring. The black breast-band is broken and narrow. The upperside is greyish-brown, the underside pure white and the legs greyish-black.

In flight. Displays well-marked white wing-bars as in Common Ringed Plover, as well as white sides to the dark-centered tail.

Female. Resembles a pale version of the male, lacking black markings on head and breast. The lore is pale or faint ochrous-buff and the broken breast-band is of variable shape and is better described as brownish patches on the sides of the breast.

Non-breeding. Appears greyish-brown without black markings. Can be sexed from mid winter when the male attains the black patterns of head and breast.

Juvenile. Has scaly greyish-brown upperside with pale fringes and pure white underside. Told from both Common and Little Ringed Plovers by more slender black bill and greyish-black legs.

Voice

The call is a short and abrupt '*vit*'.

The song is reminiscent of chatter from House Martins and is a rolling, guttural trill '*drrrrooit-drrit*' or '*draeee-drre-drre*'. The song is performed from a slow-winged, rowing song-flight over the territory.

 Adult breeding.
Male, same bird as the one stretching its wing on the opposite spread.

▶ Female. The difference between male and female is clear here, but can be less striking when paler males are involved. However, males are always more well-marked with a distinct black patch on the forehead and with a well-defined black, broken breast-band.
Note that resting waders are often hunched up and crouched, but completely change shape when stretching their necks in an alert position. 26.5. LG.

Habitat

The preferred breeding habitats are sandy beaches, salty marshlands at brackish coasts, salty lakes, salt marshes and newly established wetlands, preferably with muddy or sandy bottoms.

The same habitats are preferred outside the breeding season. Rarely seen near fresh water.

Breeding biology

Breeds when one year old and has the same mate for one or several years. Breeds singly or in loose colonies, with a good distance between nests.

The nest is a scrape in the ground which is scantily lined with small pebbles, fragments of mussel shells or plant material, preferably close to water.

The clutch of usually three eggs is incubated for 26 days by both parents. If the brood is lost, re-laying may be attempted for up to three times. In most cases the female leaves the male and the chicks shortly after hatching, and polyandry (female mating with several males) as well as polygyny (male mating with several females) may occur.

▲ ▶ Adult breeding.
Male with day-old chicks.
In the Kentish Plover, it is most often the male that leads and protects the young.
Until the chicks are able to fly, the family does not stray from the low vegetation at the edge of the beach. There they are totally camouflaged, thus mostly avoiding predation from the corvids and gulls that constitute the largest threat.
Shortly after hatching the chicks are able to feed independently. Initially the food is mostly insects, spiders and other small creatures. Later crustaceans and worms are snatched on the sandy banks.

▶ When rest is needed, or the temperature drops, the chicks of waders seek shelter and warmth under the belly feathers of the parent bird.
Here the male has acquired six tiny additional legs. Both photos 26.5. LG.

The young are able to fly after a good 30 days and become independent shortly after.

The oldest ringed bird lived for 19 years.

Migration

Arrival at the European breeding areas takes place from February to April.

Autumn migration happens soon after the breeding season, as early as from June, when the birds gather in smaller groups on suitable feeding places. In north-western Europe, birds gather particularly at the Wadden Sea, before the regular southbound migration during August–September. Birds from western Europe winter along the coast of southern Europe and western Africa.

East European birds migrate to destinations in southern Eurasia and northern Africa.

Distribution

The Kentish Plover is the least common of the three European *Charadrius* species.

The population is estimated at between 13,000 and 23,000 pairs. Denmark marks the northern limit of the breeding distribution. The nominate, *C.a. alexandrinus*, breeds from western Europe, including the Atlantic Islands and northern Africa, to the east through the Middle East and southern Asia to the Pacific Ocean. There are two further subspecies in east and south-east Asia. The largest populations in Europe are found in Portugal, Spain, Italy and Greece.

▶ Juvenile with typical scaly appearance on head and upperside as well as with fresh pale-edged tertials.
Note the four buff-fringed large scapulars, and the smaller feathers on the neck-side which indicate that the bird is moulting into its first winter plumage.
Note also the black bill, white eye-ring and greyish-black legs which sets it apart from the two more common banded plovers, Common and Little Ringed Plovers, as well as from juvenile Semipalmated Plover. 30.8. Azores. S.P.

▲ Adult non-breeding. Very similar to juvenile, but still has a visible, narrow, broken breast-band. An additional feature is the larger and less tidily placed feathers of the upperside, which have narrower or worn-off pale fringes.
Note the white sides of the rump and the pure white outermost tail feathers that distinguish it from the other European banded plovers. 12.9. Israel. LK.

KILLDEER
CHARADRIUS VOCIFERUS

Meaning of the name
'The noisy plover'
Charadrius, derived from the Greek *kharadra*, creek, relating to the plover habitat of gravelly streams and creeks.

Vociferus derived from *vocare*, shout, and from *fero*, carry/raise. Referring to the noisy behaviour of the species.

Jizz
L. 23–26 cm. Ws. 59–63 cm.

Medium-sized, long-tailed and long-legged American species with characteristic, double breast-band and prominent red orbital ring throughout the year.

In flight displays a broad, white wing-bar and conspicuous, cinnamon-coloured rump and tail pattern. Very noisy on the breeding ground.

▲ Adult non-breeding.
The long cinnamon-washed tail, the diagnostic tail-edge pattern and the prominent wing-bar make the species easily recognisable in flight. The cinnamon-coloured edges on mantle, scapulars and coverts are evidence of active moult from breeding to non-breeding. 31.8. Cape May, Florida. TH.

▼ Post juvenile is almost identical to adult non-breeding and quickly attains the red orbital ring.
Note the difference between the new, first-winter feathers with cinnamon edges and the pale, unmoulted juvenile coverts with pale buff edges.
31.10. California. DP.

Plumage and identification

The sexes are alike, though some females have some brown mixed in the black head markings.

Adult breeding. First and foremost recognisable from the European banded plovers by the double breast-band, which is present throughout the year, as well as the distinct red orbital ring and the flesh-coloured legs. The upperside is brownish-grey and has cinnamon-coloured fringes in fresh breeding plumage.

Non-breeding. Like adult breeding, but all new feathers have cinnamon fringes.

Juvenile. Very similar to adult but has yellow orbital ring.

Voice

The English name stems from the characteristic call of the species, '*klooheee*' which with some measure of imagination could sound like 'kill-deer'. This call is heard in countless variations from short, shrill single notes to longer rolling ones.

Performs a song flight which is quite similar to that of Common Ringed Plover.

Migration and distribution

Very rare vagrant from America. Recorded in the Faroe Islands, France, Hungary, Iceland, Ireland, Norway, Portugal, Spain, Sweden, Switzerland and the United Kingdom.

The nominate form has a North American distribution and two additional subspecies breed in Central and South America.

▼ Adult breeding in worn breeding plumage with a few remaining cinnamon-coloured fringes on mantle and scapulars. Characteristically long-tailed with cinnamon-coloured uppertail coverts, double breast band and diagnostic red orbital ring. 6.6. Canada. DP.

EURASIAN DOTTEREL
EUDROMIAS MORINELLUS

Meaning of the name
'The foolish, good runner'
Greek: *eu*, good and *dromos*, running.
Morus, foolish.

The Greek name of the species is probably derived from the unwary, confiding behaviour towards humans.

Jizz
L. 20–22 cm. Ws. 57–64 cm.
Similar shape to European Golden Plover, but is more plump with broad, pigeon-like breast. The pale chin, the bold, white supercilium and the white vent contrasting to the tricoloured underside in blackish-brown, rusty-red and grey reveal resting birds at quite a distance.

In flight. The flight silhouette is similar to that of the Golden Plover, but it has narrower and longer wings lacking a wing-bar. Is very confident both on and away from the breeding area.

Plumage and identification
Adult breeding. Female more stockily built and with more saturated colours than the male.

Has pale head with a short, black bill, dark speckled crown, white throat and chin, and prominent white supercilium which extends to the nape where it forms a white V.

Neck and upperside greyish with buff fringes on the scapulars and coverts.

Has a white breast-band and brick-red breast which blends into blackish-brown on the belly sharply contrasting to the white vent. The legs are yellowish.

In flight. The upperside is uniformly greyish without a wing-bar. The tail is white-tipped with a black subterminal band. The underwings are grey.

▼ Adult breeding. Presumed female due to uniform neck and upper breast and extensive dark belly patch. During the spring migration smaller groups of Dotterels are seen on the regular roosting sites. The sexes are easier to tell apart on the breeding ground when the moult has been fully completed. 18.5. NLJ.

▲ Adult breeding. Size and flight profile of a European Golden Plover, but has wider body and longer wings. The hand in particularly is longer and more narrow. 13.5. HS

▲ Adult breeding. Unmistakable at close range with conspicuous head markings and tricoloured underside in grey, reddish-brown and white. 13.5. HS.

▲ Adult breeding. The upperside is plain greyish-brown without wing-bars, but has a dark bar across the white-tipped tail. 13.5. HS.

Non-breeding. In this plumage the upperside is paler, and the striking colours of the underside have been substituted by greyish-brown and a visible but diffuse breast-band.

Juvenile. Overall pale greyish-buff with mantle and scapulars speckled in black and white, and with coverts and tertials edged pale buff.

Voice

The call is reminiscent of that of Dunlin, but is a softer, more rolling '*plyoohrr*'.

The song, which is carried out by the female in an undulating flight over the territory, is a repeated soft, melodious trill, not unlike that of a Greenfinch.

Habitat

From late February small groups on migration can be observed on regular roosting areas such as moorland, ploughed fields, mountain plateaus and recently cultivated plains. The breeding habitat is situated above the tree-line on stony, Arctic tundra and high-altitude alpine plateaus with short vegetation.

Breeding biology

The usual sexual roles are reversed in the Dotterel. The more strongly coloured female is the first to arrive on the territory where she performs the song flight while eagerly calling. The female often pairs with several males. Pair bonds last for one or more seasons, and pairs may nest in small, loose colonies. Males may also mate with two or more females.

The clutch most often consists of three eggs which are laid in a shallow scrape lined with plant material. The male is left with the responsibility of the clutch when the female, after a week's time, starts to lay a second clutch with a new male. Later she may even produce a third brood. After *c.* 26 days the eggs hatch and by then the incubating male has lost close to 8% of his body weight. The welfare of

▶ Adult breeding. Presumed female.
The female can often be told from the male by darker, more uniform crown, evenly grey-coloured neck area, and the more well-marked, white breast-band.
Note that the underwing is evenly grey (but may appear white in strong light), as opposed to the flashing, whitish underwing of the European Golden Plover. 13.5. HS.

the young is mainly the concern of the male since the majority of the females leave the territory 2–4 weeks before the rest of the family. However, some females help with incubation and care of the young. They feed on insects, spiders, other small creatures and berries, later augmented by worms during the autumn migration. The male accompanies the young until they are able to fly after a good 25 days, and on the southbound migration.

Migration

In spring the migration crosses Europe in a broad front and the birds stop to rest at regular stop-off locations, such as in the Alps.

The breeding areas are reached from mid-May to mid-June. The return migration towards the winter quarters is initiated in July and August by the adult females, followed by the males and the juveniles in August–September. The autumn migration often takes place in a non-stop flight to the winter quarters in northern Africa through the Middle East and the Arabian Gulf to western Iran. Birds from eastern Siberia are believed to execute the 10,000-kilometre journey to the winter quarters in one uninterrupted flight.

Distribution

The European population counts between 12,800 and 48,400 pairs. The main distribution in Europe stretches from northern Britain, primarily the Scottish highlands, through the mountainous areas of Scandinavia and Finland into northern Russia.

The largest populations are found in Russia, Norway, Sweden, Finland and in the British Isles. There are also some other minor and irregular populations, for example in the Pyrenees and the Alps.

East of the Urals the distribution stretches along the northern coasts of Siberia to the Bering Strait as well as in southern central Asia, Kazakhstan, Siberia and Mongolia.

 First winter.
Most of the wing coverts are obscured by the breast feathers, but note the fresh scapulars with large, dark feather-centres and the new, upper tertials with cinnamon-coloured edges in contrast to the underlying, worn, pale-edged juvenile ones. 5.1. Oman. HJE.

◀ Adult breeding.
Male on the breeding ground with day-old chicks. The male can be told from the female by the streaked crown and throat area as well as on weaker coloration and less extensive brownish-black belly-patch. 6.7. Norway. DP.

▲ The species as a rule breeds when two years old, but some individuals do so when one year old, and almost all young individuals attain the breeding plumage.
This bird is presumably an adult non-breeding to breeding but may be a first summer, which can be told from adults on retained inner medium arm coverts, which are often hidden by the scapulars, and on the very worn flight feathers. 6.5. HS.

 Juvenile.
Differs from adults by the overall buffish and scaly look with white 'pearls' at the sides of the dark tips of the scapulars and on the median arm coverts, as well as by buff edges to the tertials.
May be confused with juveniles of the golden plover species, but can always be told from these by the distinct, whitish-buff supercilium, the white, horizontal breast-band and the buff flanks. 8.8. JL.

CASPIAN PLOVER
CHARADRIUS ASIATICUS

Meaning of the name

'The Asian plover'

Charadrius, name of bird derived from the Greek *kharadra*, creek, relating to the bird's habitat of gravelly streams in creeks.

Asiaticus, Asia.

Jizz

L.18–20 cm. Ws. 55–61 cm.

Slender, elegant and long-billed banded plover with an upright stance. Large eyes and pale head where the white forehead adjoins the broad, white supercilium which stretches onto the nape. The rather long bill is black.

Has broad, greyish-brown or rusty-red breast-band. The rest of the underside is white, and the legs are greyish. The long legs and long bill give the impression of a larger bird, but the body size is the same as that of Common Ringed Plover.

In flight. A short, white wing-bar is noticeable in the centre of the wing.

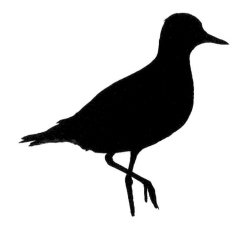

Plumage and identification

Adult breeding, male. Has white head with greyish-brown crown and eye-stripe that join in the nape merging with the greyish-brown upperside which has thin, pale fringes in fresh plumage. The bill is black and lapwing-like.

The broad breast-band is of a beautiful chestnut colour and is bordered by a thin, black line at the white belly. The legs are greyish to greyish-green.

Female. As male, but with a buff wash to the white head markings and with greyish-brown breast-band.

In flight. The upperside is dark brown in contrast to the black hand. Has a short, white wing-bar at the inner primaries and the outer secondaries. The underwing is pale greyish-white.

Adult non-breeding. Both sexes as female breeding, but with faded, greyish-brown breast-band. Difficult to distinguish from first winter. Refer to photos and captions.

Juvenile. In fresh plumage has obvious pale cinnamon fringes on the brown upperside, becoming paler buff on the coverts. Yellowish legs. Otherwise very much like non-breeding.

Similar species

Male and female in breeding plumage are very similar to the long-legged Asian species, Oriental Plover, *Charadrius veredus*, of which the male can have an almost completely white head.

This species lacks the wing-bar and is uniform brown above. Oriental Plover has only once been recorded within the Western Palearctic (in Finland in 2003), and is therefore not described in this book.

◀ Adult breeding, female.
The female in breeding plumage is a pale version of the male. In adult non-breeding the sexes are similar, and are difficult to tell from first winter birds. 10.4. Israel. KBJ.

▼ Adult breeding, male.
In breeding plumage the male is unmistakeable among the European banded plovers with its predominantly white head and the broad, chestnut breast-band edged black on the lower breast. 10.4. Israel. KBJ.

Voice

The call is a short whipping '*choohp*', which may be repeated in a longer series.

Performs song flight over the territory uttering a melodious, ringing '*tew-urrlee, tew-urrlee*'

Habitat

The breeding habitat is semi-desert and saline steppe with sparse vegetation, preferably near water. Very rare and irregular vagrant from south-eastern Russia and Kazakhstan. Uses similar habitat in the winter quarters.

Breeding biology

The species is little-studied.

It is believed to breed as a two-year-old. As is usual for banded plovers, its nest is a shallow depression lined with plant material and other suitable matter. The single clutch of three eggs is incubated by both parents for presumably *c*. 25 days and the young fledge after *c*. 30 days.

After the breeding season, family flocks gather on riverbanks and wetlands.

Migration

Arrival to the breeding areas takes place from late March to April after migration stop-offs, for example, in Israel.

Leaves the breeding grounds in August to October, with stop-offs in Iran, Iraq, on the Arabian peninsula and at the Red Sea, before the winter quarters are reached in the north-eastern, eastern and southern parts of Africa.

Distribution

The breeding distribution stretches from southern Russia around the Caspian Sea down to Turkmenistan and through Kazakhstan to north-western China.

◀ Juvenile.
Differs from non-breeding in having a scaly crown due to buff feather fringes present on the entire upperside. Most conspicuous are the broad, richer coloured edges of mantle and scapulars. There is also some contrast to the more sandy fringes of the coverts, but these traits are only reliable in fresh plumage. 12.9. Oman. HJE.

▶ Adult male.
Non-breeding to breeding.
Determined by fresh tail and wing feathers and emerging chestnut breast-band.
   Note the overall pale greyish-brown underwing with pale brown axillaries, vaguely contrasting to the whitish and blackish-brown lesser coverts. Oman. 30.3. BLC.

▶ First winter, determined by the fresh plumage where juvenile scapulars with cinnamon-coloured edges can be seen.
   Note also the short, pale wing-bar formed by the pale-tipped outer arm coverts and the pale bases to the inner primaries. The white shaft of the outer primary is evident as well, something that also applies to Eurasian Dotterel but that species has a uniform, greyish-brown upperside without white wing-bar. 28.8. Oman. HJE.

◀ First winter with retained juvenile coverts and tertials. 28.8. Oman. HJE.

LESSER SAND PLOVER
CHARADRIUS MONGOLUS

Meaning of the name

'The plover from Mongolia'
Charadrius, name for plovers, derived from the Greek *kharadra*, creek, relating to the bird's habitat along gravelly streams in creeks.
Mongolus, Mongolia.

Jizz

L.18–21 cm. Ws. 45–58.
Medium-sized banded plover with short neck and a round head. The position of the eye is not central but slightly nearer to the bill. Resting birds may seem slightly unbalanced due to the legs being positioned just behind the centre of the body. An important character when trying to separate Lesser Sand Plover from Greater Sand Plover is the former's saddle-like shape of the upper mandible, with a slight depression near the middle, and an oblong bulge to the bent tip of the bill. The latter generally has a heavier, dagger-like bill with just a hint of saddle shape, and a more evenly tapering

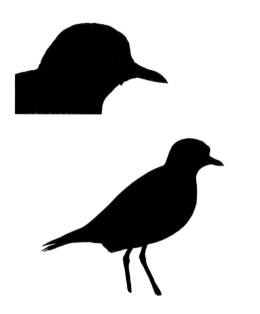

▲ Male, *atrifons* group, with black 'bandit mask' and with no narrow black upper line to the breast-band. The feet may either project beyond the tail or not, depending on the variable length of the legs between the subspecies. 28.4. Khok Kham, Thailand. HS.

▲ Non-breeding. Note the white wing-bar which is of even width on the arm, but broader and shorter than that of Greater Sand Plover on the hand. 9.3. Khok Kham, Thailand. HS.

tip. Note that the smallest subspecies of Greater Sand Plover, *Charadrius leschenaultii columbinus*, has a finer bill, but with the same shape of the tip.

The general rule of thumb when separating Lesser Sand Plover from Greater Sand Plover is that if in doubt, it is usually a Lesser Sand Plover.

Plumages and key features

Adult breeding, male. Has greyish-brown upperside and head with either an all-black 'bandit-mask', or a white blaze separated by a narrow black frontal bar, depending on the subspecies. Frontal part of the crown, nape, upper breast and flanks are orange-red to deep ochreous in contrast to the white throat and belly. Depending on the subspecies, the red breast may be bordered by a thin, black line at the neck. Belly beige-white, legs back to greyish-black.

Female. Has ill-defined 'bandit-mask' of a greyish or brownish-grey colour (sometimes absent) and appears paler with weaker and less extensive markings on the neck and breast area.

Non-breeding. See photo text.
Juvenile. See photo text.

Subspecies

The species has been divided into two groups with a total of five subspecies. The south-western *atrifons* group, including atrifons, pamirensis and schaeferi, as a rule has all-black 'bandit-mask', and no black line to the breast-band. The *mongolus group* with the nominate, mongolus, and stegmanni, has the white blaze divided by a black frontal bar and a narrow black line between the reddish breast and the white throat.

A safe identification of the individual subspecies is difficult as more or less black or white can be present in the head patterns of both groups.

Voice

Call and flight call is a short and somewhat metallic '*tlllp*', very similar to the call of Greater Sand Plover.

Habitat

The species breeds above the tree line on stony, sparsely vegetated mountain plains and on tundra, preferably near water.

Outside the breeding season seen at coasts, tidal deltas and river mouths.

Migration and distribution

Rare vagrant to Europe from Asia. The *atrifons* group has a central Asian distribution, primarily in China and Tibet, while the *mongolus* group is north Asian with its main distribution in Siberia eastwards to Kamchatka.

Winters in specific groupings in areas from eastern Africa to southern Asia and to Australia.

▶ Adult breeding, female.
Typical representative of the *mongolus* group, with the white blaze divided by a narrow black line from the upper mandible to the back frontal bar. Females generally have the black facial markings admixed with brown and grey, and are less bright in the coloured areas of the crown, nape and breast than the males.

Note that the ochreous-red breast is bordered at the white neck by an ill-defined and very thin black collar. This bird was photographed at the breeding ground on Chukotka in north-eastern Siberia. 21.6. GV.

▶ Adult breeding.
Typical male from the *atrifons* group with an all-black 'bandit mask' and no white above the bill.

Males of all subspecies as a rule have wholly black facial markings and richer orange-red coloration than the females. Note the bill shape of this male, with just a mere hint of a bulge on the outer third of the otherwise straight culmen. The bill still appears more petite with a more pinched-off tip than that of the similar Greater Sand Plover. 28.4. Khok Kham, Thailand. HS.

▶ First winter. The *atrifons* group.
Age determined by the retained juvenile coverts with broad, buff fringes.

Note that juveniles often have greenish legs.

Adult non-breeding has thin pale fringes on the upperside and more well-defined breast-band, which may be unbroken in the middle. Furthermore it has a paler forehead and supercilium. 23.10. Oman. AA.

GREATER SAND PLOVER
CHARADRIUS LESCHENAULTII

Meaning of the name
'Leschenault's plover'

Charadrius, name for plovers, derived from the Greek *kharadra*, creek, relating to the bird's habitat along gravelly streams in creeks.

Leschenaultii, named after Jean Baptiste Louis Claude Théodore Leschenault de la Tour, French botanist and ornithologist (1773–1826).

Jizz
L. 22–25 cm. Ws. 53–60 cm.

Slightly larger and more robust than the Lesser Sand Plover with a more powerful appearance due to its flatter crown and a heavier, more pointed bill. The bird seems more symmetrical with the eye placed in the centre of the head, and the legs placed exactly below the centre of the body. The shape of the body bears resemblance to a small Grey Plover. The shape of the bill is symmetrical with an evenly attenuated tip, lacking the saddle-like bulge and

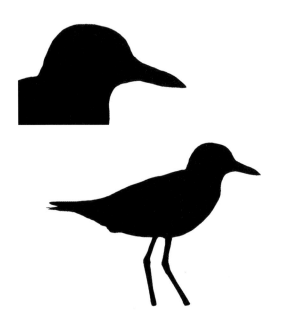

▲ Adult breeding, male. In flight very similar to Lesser Sand Plover, but has a more flattened head, longer and more robust bill as well as more white on the sides of the tail. The feet may or may not project beyond the tail, depending on subspecies as they have different lengths of legs. 16.2. Oman. HJE.

▲ Adult non-breeding to breeding with emerging ochreous feathers on breast-band and ear-coverts. Note the long, white parts of the inner primaries that form an almost rectangular 'speculum', which is larger and more clear-cut than on Lesser Sand Plover. 16.2. Oman. HJE.

pinched-off tip of the upper mandible seen in Lesser Sand Plover.

The subspecies *colombinus* has a similar shape of bill to Lesser Sand Plover, but always has a symmetrically attenuated tip of the bill.

The general rule of thumb when separating Lesser Sand Plover from Greater Sand Plover is that if you're in doubt, it is usually a Lesser Sand Plover.

Plumage and identification
Adult breeding, male. Very similar to Lesser Sand Plover with a black 'bandit mask', often with variably extensive white blaze on the forehead, but never divided by a vertical, narrow, black frontal bar. The upperside is greyish-brown, with or without an orange tinge depending on the subspecies. Legs greyish-green to yellowish-green.

Female. Paler, often with the black head markings more or less substituted by greyish-brown.

Non-breeding. Refer to photo caption.

Juvenile. Refer to photo caption.

Subspecies
The species is divided into three subspecies; *colombinus*, *scythicus* and *leschenaultii*. Subspecies colombinus is the smallest and the one with a bill-shape most similar to that of Lesser Sand Plover, and which furthermore has the richest ochreous tinge to the upperside. Subspecies *scythicus* resembles the nominate *leschenaultii*, which has the longest and most heavy bill, and the most uniform, greyish-brown upperside.

Voice
Call and flight call is a short, soft trill '*plllp*', almost identical to the call of Lesser Sand Plover.

Habitat
A rare vagrant from the Middle East and Asia to Europe, more often recorded than Lesser Sand Plover. Breeds in dry, desert-like areas and on saline steppe. Outside the breeding season at coasts, rivers and muddy lagoons.

Migration and distribution

The subspecies *colombinus* breeds in the Middle East, Turkey and further on to Azerbaijan, where it is replaced by *scythicus,* and later *leschenaultii* onwards through central Asia and Mongolia.

The main winter quarters are the south-eastern Mediterranean, the Bay of Aden and the Red Sea, north-eastern and eastern Africa and further eastwards to western India, southern Asia and Australia, with more easterly breeders having more easterly wintering grounds.

▲▶ Adult breeding, male.
This bird's robust bill with straight culmen rules out the Middle Eastern and Asia Minor subspecies *colombinus,* which also typically has more white on the forehead. It presumably belongs to **scythicus**, from east of the Caspian Sea to Afghanistan. Blåvandshuk. Denmark. 6.6. EFH.

▶ Adult breeding, female.
Told from the male by the pale breast-band and the incomplete black head markings, primarily substituted by brownish-buff below and behind the eye.
   The powerful, long bill and the stance, which hints at a Grey Plover, as well as the locality, all indicate the nominate *leschenaultii*. 2.5. China. DP.

▶ Juvenile, *leschenaultii*.
Lesser Sand Plover and Greater Sand Plover are almost identical in the juvenile and first winter plumages.
   Juveniles of both species have greenish-yellow legs.
   Identifying these two very similar banded plovers must rely closely on the jizz, shape of bill, and placement of the eyes and the legs, as well as the location of the observation, to ensure you reach the correct conclusion. This bird was identified by the central placement of the eye, the shape of the bill, and the well-balanced position of the legs. 27.8. China. DP.

The lapwing genus, *Vanellus*, numbers 24 species, distributed on all continents except Antarctica.

The lapwings make up an easily recognised group of waders with conspicuous, often pied plumage.

Standing lapwings often adopt a strutting posture, and in flight they are unmistakeable and easily told from their relatives by their broad wings and deep-flapping, butterfly-like flight, often with twists and turns.

There are four European species of lapwings, three of which have an eastern distribution. Among them is the Sociable Lapwing which is classed as Critically Endangered with a world population of only 11,000 adult birds.

Only the Northern Lapwing is widely distributed from west to east, with an estimated European population of 1.2 million birds.

A flock of Northern Lapwings is a common sight of late summer and early autumn, when the birds roam the countryside looking for food and places to roost for the night.

The Northern Lapwing is easy to recognise when it flashes its white underside contrasting to the dark upperside, and not least by its broad wings and flapping, butterfly-like mode of flight.

This flock consists primarily of non-breeding adults, with a few juveniles. 30.10. LG.

NORTHERN LAPWING
VANELLUS VANELLUS

Meaning of the name
'The small fan'
From *Vannus*, fan and *ellus*, small.
    Named after the broad, fan-like wings.

Jizz
L. 28–31 cm. Ws. 82–87 cm.
Medium-large wader with long wispy crest and a conspicuously pied plumage. Like the other plovers it has large dark eyes and a proudly erect stance.

In flight. In flight very broad-winged with visible 'fingers' and fluttering, butterfly-like wingbeats. Noisy and conspicuous in the spring, when the male establishes his territory in the buoyant display flight.

Seen in larger flocks outside the breeding season, often on stubble fields in the company of Eurasian Golden Plovers.

▲ Adult breeding, male.
The flying Northern Lapwing is unmistakeable with the white underside, the ochreous undertail coverts, and the white wing-patches which are visible from far away. 15.4. HS.

▼ Adult breeding, male.
Differs from the breeding female by the longer crest, the all-black neck and breast and the bright reddish legs. 30.5. LG.

▼ Adult breeding, female.
Differs from the breeding male by the shorter crest (particularly abraded on this female), and by having the black breast speckled with whitish feathers. The legs are dull reddish. 12.6. LG.

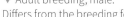

Plumage and identification

Adult breeding, male. Has a black crown which ends in a very long crest, and white head and nape with a black face and throat, which adjoins the all-black breast-bib. The upperside seems black at a distance, but is iridescent in green and violet-metallic on the median coverts. The belly is white and the undertail coverts ochreous. Reddish legs.

In flight. Very broad-winged with a rounded hand with visible 'fingers'. The upperside is dark apart from the white tail, which has a broad, black subterminal band. The underside is strikingly black-and white, and even at a great distances the white arm coverts are eye-catching.

Female. As the male but with a shorter crest, pale-speckled breast and dull reddish legs.

Adult non-breeding. Attains a less contrasting head, a shorter crest, paler legs and yellowish fringes on the feathers of the upperside. The pale fringes wear off during winter and early spring.

Juvenile. Has brownish breast-band and thin, buff fringes on the upperside, which lacks the prominent greenish sheen of the adults.

Voice

Call and alarm call is a plaintive '*wiee-eech*' with variable intonation.

The song is a characteristic whooping '*keeuvit-vit-vit-keeuvit*' intermixed with muffled wheezing and throbbing from the wings when the male performs abrubt dives during the acrobatic rolling and tumbling display flight, often low over the ground.

Adult breeding.
Male in display flight over the territory. Note the broad, rounded hand and the very long crest.
The female has browner primaries, a narrower hand and a shorter crest. 1.4. JL.

Habitat

Found in open, sparsely vegetated country, most often in agricultural areas, but also meadows, grassy plains, mountainous moors and larger stretches of boggy land. Outside the breeding season seen in larger flocks, both at coasts and inland on ploughed and stubble fields.

Breeding biology

The male often returns to the same territory for years. The pair stays together throughout the season, and breeds away from other pairs, while vigorously defending the territory.

During the breeding season the unique song flight is performed both day and night. Polygamy is not unknown, and often both sexes will mate opportunistically with their neighbours. The single clutch of four eggs is laid in a scrape in the ground, lined with plant material. The eggs are incubated by both parents for a good 25 days. The female is responsible for most parental care. The food consists of insects, other small creatures and worms. The young fledge after some 40 days.

The oldest ringed bird lived for more than 24 years.

Migration

The migration of the Northern Lapwing is strongly governed by the weather. During mild weather, birds may winter at northern latitudes or arrive very early to the breeding areas, which may result in mass starvation if the weather suddenly deteriorates.

The Northern Lapwing arrives at its breeding grounds from January to May. The males are the first to arrive in the territory. As soon as possible after breeding, most often in June, the breeding areas are deserted and during the late summer large flocks assemble on stubble and ploughed fields and at shallow coasts.

During September–November birds from Scandinavia and northern Europe migrate towards the main winter quarters in the British Isles and along the coasts of south-western Europe where they join the partly sedentary populations of north-western Europe.

Birds from southern Europe are primarily sedentary, while the eastern populations migrate to the west and southwest to the Mediterranean, north Africa, the Middle East and the Arabian Gulf.

Distribution

About 2 million pairs breed in Europe. The Northern Lapwing has been classified as Near Threatened, because of habitat loss through agricultural intensification, drainage and pollution, but it still breeds in most of Europe eastwards through Turkey, Iran, Kazakhstan, Russia, southern and eastern Siberia, Mongolia and northern China to the Pacific. The largest populations live in eastern Europe, Finland, Sweden, Germany, the Netherlands and France.

The juvenile differs from the adult by having brownish head markings and breast-band as well as buff fringes on the neatly arranged juvenile feathers of the mantle, scapulars and coverts. 9.8. JL.

Grey Plover and Northern Lapwing.
Grey Plover, female, adult breeding. Sexed on blackish-brown markings mixed with white feathers on breast and belly. The male is all black from chin to belly. Note the blackish axillaries (armpit), which is a unique feature of the species. Northern Lapwing, adult non-breeding. Determined on the golden fringes of the larger feathers of the mantle, scapulars and coverts which are arranged in a somewhat irregular fashion, contrary to the neat rows of scaly feathers in the more orderly plumage of a juvenile. The two species fight over a morsel of food, or over an attractive feeding territory on the beach. Here the Northern Lapwing holds its grounds and is not deterred by the aggressive Grey Plover. 26.8. JL.

SPUR-WINGED LAPWING
VANELLUS SPINOSUS

Meaning of the name
'The small fan with thorns'
From *vannus*, fan and *ellus*, small. *Spinosus*, thorny.
Named after the thorny spurs on the carpal joint.

Jizz
L. 25–28 cm. Ws. 69–81 cm.
A very long-legged and slender, brown, white and black lapwing. The only European lapwing with a red eye. Is aggressive both during and outside the breeding season, particularly towards other waders. Forages with calm movements, taking a few steps at a time.

In flight. Flies in a typical lapwing manner and the black secondaries distinguishes the species from the similar Sociable Lapwing which has white secondaries.

The toes and almost half of the tarsus extend beyond the tail in flight.

▲ Adult breeding.
Presumed male due to the length of the spurs on the carpal joint. The spurs, which give the species its name, measure up to 12 mm in males, and are thinner and more curved than those of the female which are shorter and thicker, measuring up to 8 mm.
   The spurs come in handy in the defence against predators during the breeding season, and against rivals in the mating season. 25.3. Israel. EFH.

▼ Adult breeding.
The sexes are alike throughout the year and are unmistakeable among the waders of Europe with the tricoloured plumage of brown, black and white.
   In a more upright stance, the comparatively long neck and a short, black crest on the nape are noticeable. 21.4. Israel. KBJ.

Similar species

May resemble Sociable Lapwing in flight, but that species has white secondaries, whereas the Spur-winged Lapwing has black secondaries.

Plumage and identification

Adult. The sexes are alike throughout the year. The male can often be told from the female by its longer and more curved spurs. The upperside is brownish-buff with long, lanceolated scapulars and the head is black with a red eye. The cheek, sides of the neck and the neck are white, separated by a black chin-stripe which adjoins the black breast. Black legs.

In flight. The upperside is brownish-buff with a white bar separating the brown arm from the black primaries. The tail is white with a broad, black sub-terminal band. The underside is black and white.

Juvenile. Until late autumn juveniles differ from the adults by having less extensive black markings, and buffish fringes to the greyish-brown upperside.

Voice

Noisy and vocal in the breeding season, uttering loud and quickly repeated '*kick-kick-kick*'.

More high-pitched and shrill when agitated.

Habitat

The breeding habitat is dry, stony or gravelly areas near rivers, lakes and ponds, but also in wetlands and on cultivated fields.

Breeding biology

Aggressive and strongly territorial, particularly against other waders. However, conspecifics are tolerated within the territory, which may be held throughout the year.

The pairs breed singly or in loose colonies. The 2–4 eggs, occasionally five, are laid in a shallow scrape, unlined, or lined with plant matter or pebbles. The clutch is incubated by both sexes for *c.* 23 days, and the young are defended and cared for by both parents until they are ready to fly after 6–8 weeks. In the case of an additional brood, the male takes over the responsibility for the first one.

Migration

Greek and Turkish breeding birds arrive in March and depart in the course of October.

The migration passes over Crete and Cyprus towards the Middle East, the Arabian Peninsula and Africa.

Distribution

A small population of between 70 and 120 pairs breed on Cyprus and in north-eastern Greece. The species is a rare vagrant to the rest of western Europe. The Spur-winged Lapwing is distributed from north-eastern Greece through Turkey to the Middle East, and also occurs in Africa in the delta of the Nile and along the equator.

An adult female, breeding, with short, thick spurs is chasing a young Black-winged Stilt out of her territory. The Black-winged Stilt is a first summer, determined by the retained pale tips of the inner primaries. 29.4. KBJ.

SOCIABLE LAPWING
VANELLUS GREGARIUS

Meaning of the name
'The sociable little fan'
From *vannus*, fan and *ellus*, little.

Grex, *gregis*, flock. Refers to the loose colonies in the breeding areas, and the sociable behaviour during migration.

Jizz
L. 27–30 cm. Ws. 70–76 cm.

Of medium size as the other lapwings, with medium long legs, erect posture and cautious behaviour. Employs the characteristic lapwing manner of feeding by running, stopping and snatching.

In flight. Has conspicuous tricoloured upperside with white secondaries, not black as in the similar Spur-winged Lapwing. Only a little of the toes project beyond the tail in flight.

Plumage and identification
Adult breeding. Has grey upperside and a black crown which is separated from the buff head by a white forehead and a broad, white supercilium which extends to the nape. The bill is black and straight. The lore and a narrow eye-stripe behind the eye are dark.

On the underside the grey breast fades into the black belly patch, turning chestnut on the lower belly.

The vent is white and the legs are black.

In flight. Has grey back and front of the wing in contrast to the black hand and a wide, white marking along the secondaries. The tail is white with a rectangular, black band bordered by white.

Female. In the breeding plumage some females can be told from the males by brownish-grey belly with scattered pale feathers.

Adult non-breeding. Greyish-brown without dark belly. Pale fringes to the feathers of the upperside and streaked crown, neck and upper breast.

▲ Adult non-breeding in developing breeding plumage with increasingly grey upperside. Note the streaked upperbreast which can just be seen. The white secondaries set the species apart from the Spur-winged Lapwing, which is also of south-eastern European distribution. The latter has all-black secondaries. 18.1. Oman. HJE.

◀ Adult breeding.
The bird is almost in full breeding plumage with only a few streaked feathers remaining of the winter plumage on the crown, neck and upper breast, among the emerging grey feathers. 21.1. Oman. ISA.

Juvenile. Resembles adult, but does not have a dark lore. The feathers on the upperside have distinct, pale buff fringes, and the neck and upper breast are densely streaked with V-shaped markings.

Voice

The call is a dry, grating '*creet*', which resembles something between a croaking frog and a Corncrake. It is also heard in almost snarling series from agitated birds in the breeding territory.

Habitat

Migrants occur in open habitats, often in company of Northern Lapwing. Breeds on grassy and mugwort steppe and on cultivated land.

Breeding biology

At the beginning of the breeding period small groups of males are seen performing their aerial display to impress their future partners. The pair is monogamous and breeds in loose colonies. The 2–5 eggs are laid in a shallow depression lined with plant matter or pebbles. The single clutch is incubated for 22 days, primarily by the female. The young fledge after 35–40 days.

This handsome lapwing is unfortunately highly threatened by cultivation of the steppes, and the trampling of eggs and young by domestic animals such as goats, sheep and cattle, as well as by predation from foxes.

However, the major threat seems to be hunting on the migration route and in the winter quarters.

Migration

Arrives to the breeding areas in April–May and departs in September–November towards the winter quarters in the Middle East, north Africa, the Arabian peninsula and in north-western India and Pakistan.

Distribution

Scarce migrant to south-eastern Europe and rare vagrant to western Europe. Breeds within Europe, with 0–10 pairs in south-eastern Russia. The rest of the world population of a good 11,000 adult birds primarily breeds in Kazakhstan.

▼ The juvenile differs from the adult in winter plumage by the lack of a dark lore, and by having the dark feathers of the upperside distinctly edged buff. It also has a prominently streaked neck and spotted breast with V-shaped marks. 4.11. Oman. HJE.

▼ Adult non-breeding. Note the hint of a black stripe between the bill and the eye, uniform pale greyish-brown upperside with narrow pale fringes and leaf-like, brown markings on the breast. The few pale grey feathers are the first signs of the future breeding plumage. 21.1. Oman. ISA.

WHITE-TAILED LAPWING
VANELLUS LEUCURUS

Meaning of the name
'The white-tailed small fan'
From *vannus*, fan and *ellus*, small. Greek, *leuko*, white, *oura*, tail. After the white tail.

Jizz
L. 26–29 cm. Ws. 67–70 cm.
A greyish-brown lapwing with stout, black and relatively long bill and with very long, yellow legs.

In flight it displays an all-white tail, a unique feature among the European lapwings.

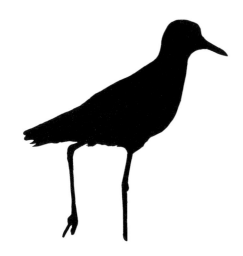

Plumage and identification
Adult. The sexes are similar, and the adult birds have the same plumage throughout the year.

The upperside is pale greyish-buff. The breast is grey and the belly rosy.

In flight. Has a broad, white diagonal bar between the brown back and forewing, and the black hand. The tail is completely white, as opposed to Sociable Lapwing and Spur-winged Lapwing which both have black-and-white tails.

Juvenile. Has strongly black-speckled upperside.

Voice
The alarm call is a squeaky '*wee-eck*'. The same call is given within groups of several birds and becomes a squeaky and plaintive song, but rapidly repeated and more drawn-out; '*weuuk-weuuk*'.

Habitat
Breeds at shores of lakes, rivers, brackish lagoons and marshes, preferably with ample vegetation and close to calm water where it can wade after insects, small creatures and fish.

Also forages on damp grassland and in rice paddies. Occurs in a similar habitat outside the breeding season.

Breeding biology

Monogamous and may breed singly, but often nests in loose colonies where territories may be shared with pratincoles and Black-winged Stilts. The four eggs of the single clutch are laid in a sparsely lined, shallow depression in the ground. Incubation lasts for 22 days, and the young are able to fly after *c.* 30 days. They thereafter remain in the area with the parents.

Distribution and migration

A rare vagrant to western Europe. Breeds locally in Turkey, Azerbaijan and the Middle East with the main distribution east of the Caspian Sea to Eastern Kazakhstan. The migrating populations winter in Sudan, around the Arabian Gulf, in Pakistan and in India.

▲ Adult breeding. An unmistakable lapwing on the European continent. 1.5. Dubai. HJE.

▲ Adult breeding. Note the all-white tail, which sets it apart from similar species of lapwings. 1.5. Dubai. HJE.

◀ Adult breeding.
The White-tailed Lapwing has the same plumage throughout the year. The long, yellow legs are a striking feature. The species is a rare vagrant to Europe, but it has bred on Cyprus and in Romania. 30.3. Dubai. HS.

▶ Post-juvenile.
Juvenile in fresh plumage has evenly patterned upperside with black subterminal bands and pale buff and whitish fringes. This bird is in active moult into first winter and is rapidly becoming similar to the adults. 7.11. Oman. ISA.

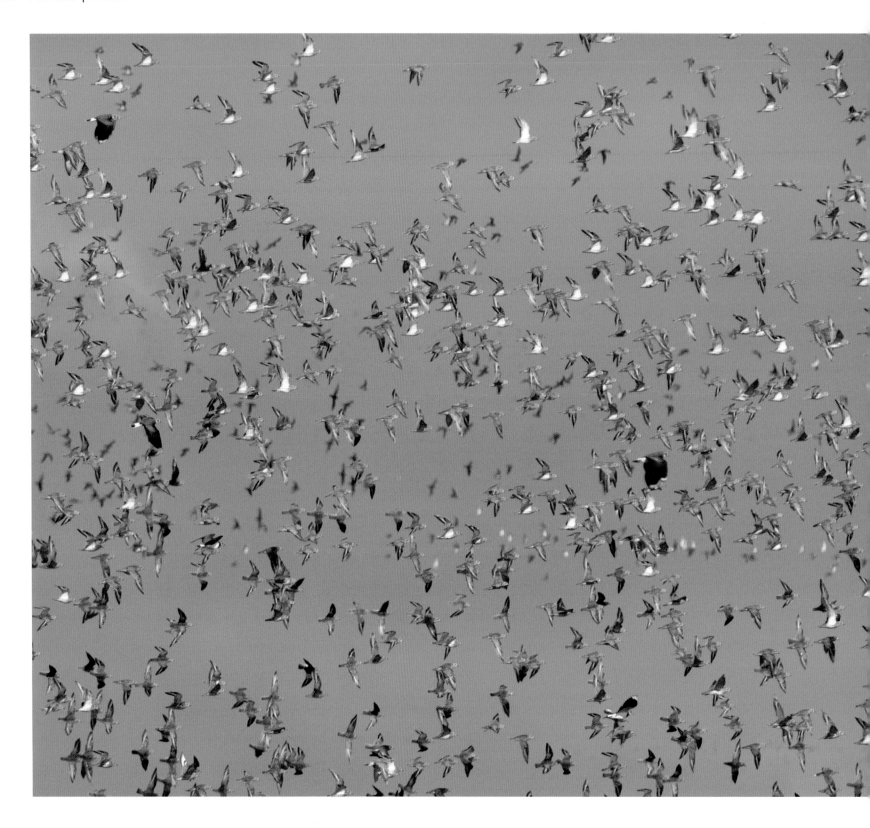

A large flock of European Golden Plovers and a few Northern Lapwings, swirling around in an attempt to confuse a passing Peregrine.

The medium-sized European Golden Plover typically gathers in large flocks during the autumn migration, as here in the Wadden Sea, where identification of the flock can be established at a long distance by the brown uppersides with white patches on the hand, and the almost sparkly flashing of white underwings, when the birds simultaneously turn in the air. 1.10. LG.

The genus *Pluvialis* is a small and easily recognisable group of plovers, with just four species, which all breed in the Northern Hemisphere.

Two species breed in Europe, while the other two are rare vagrants, one from North America and the other from east Asia.

The *Pluvialis* plovers are similar in size to the lapwings, but more stocky and evenly proportioned.

The common denominator is the chequered upperside with, in the breeding season, a black underside which extends from the black face to the belly.

Their behaviour is cautiously hesitant, and the large eyes together with the mournful call contributes to a somewhat sorrowful overall impression that distinguish these plovers from all other waders.

EUROPEAN GOLDEN PLOVER
PLUVIALIS APRICARIA

Meaning of the name
'The rain-calling lover of the sun'
From *pluvia*, rain. Related to the autumn migration which heralds the rainy season as well as the behaviour of the species; when roosting birds huddle together and 'call for rain' with their mournful whistles.

Apricari, to warm oneself in the sun, derived from *apricus*, sun-drenched, and *aperire*, to expose, to show.

The contradictory name of the species in relation to the name of the genus possibly stems from observations of European Golden Plovers which have been drying themselves after a heavy shower.

Jizz
L. 26–29 cm. Ws. 67–76 cm.
The European Golden Plover is the largest and most robust of the three species of 'golden' plovers, but the shortest-legged one.

The length of the tibia is noticeably shorter than in the other two species.

In flight. The flight is powerful and straight, showing a broad arm and a pointed falcon-like hand. Often seen descending in large circles from

the migration height to the roosting areas. The underwings are very pale. The toes do not project beyond the tail.

Plumage and identification
Adult breeding, male. The crown, back of the neck and the upperside are flecked with black, white and golden. The head, neck, breast and belly are more or less black or blackish-grey, but are separated along the crown, nape and flanks by a white stripe. The vent and undertail coverts are white, sometimes with an all-black or white-speckled wedge extending to the tip of the tail. This wedge is elongated by the black legs.

▲ Adult breeding, male.
In full breeding plumage the male of the northern subspecies, *altifrons*, has a black head and underside as opposed to the nominate southern subspecies, which often has a paler, greyish-black head and a more narrow, black breast-band.

Note that the axillaries and the inner part of the arm are white, distinguishing it from a long distance from the Grey Plover which has black axillaries.

The two rare vagrants, American and Pacific Golden Plovers, both have greyish-brown underwings. Iceland. 23.5. JL.

▼ Adult breeding, male and female respectively of the northern subspecies, *altifrons*. Both the sexes and the races overlap quite a bit in plumage, hence a safe identification must rely on the geographical location, and on direct comparisons of the pair. Iceland. 12.5. JL.

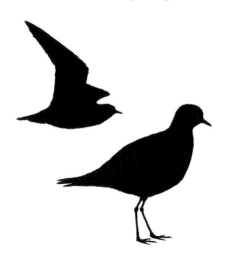

In flight, upperside. Has a pale wing-bar along the greater coverts which widens on the pale wanes of the primaries.

In flight, underside. The axillaries and the median and lesser coverts are white, which distinguishes the species from the other golden plovers.

Female. As male, but with blackish-brown or completely pale head markings and with brownish-black neck and breast pattern interspersed with white feathers. Note that males of the southern subspecies, apricaria, are paler with less extensive black on underside and have greyish-black head and neck markings thus resembling females of the northern form, altifrons. The two subspecies overlap within their respective geographical distribution.

Adult non-breeding. Has a mainly golden look due to the golden-edged dark feathers of the upperside. The breast and the flanks are streaked in the same shades. The belly is white.

Juvenile. Has an overall golden look with white and golden feather-edges of the upperside. The underside is greyish-white speckled with dark grey.

First winter. Almost identical to adult, but until mid-winter possible to distinguish by juvenile tertials, greyish-white belly and some retained juvenile coverts.

Subspecies

The species was formerly divided in northern and southern variants, but are now recognised as two subspecies. The northern, *P. a. altifrons*, which is more black on head, breast and belly, is found from eastern Greenland, Iceland and the Faroe Islands through Fennoscandia and Russia to the Taymyr.

The southern nominate form, *P. a. apricaria*, which is generally paler, breeds in the northern British Isles, sporadically in western Germany, southern Scandinavia and in the Baltic.

▲ Adult breeding to non-breeding, on autumn migration in different plumages and stages of moult. At the bottom left, a male with emerging inner primaries, and old, worn P7–10. The other individuals have still not moulted the worn, outer P10. Notice that the top left bird has black undertail coverts, as has the top bird on page 162, and this may point to American and Pacific Golden Plovers. But do remember that the two latter species have greyish-brown (not whitish) axillaries and arm coverts of the underwing. 3.9. HS.

▼ Juvenile, standing. Differs from the similar, but more powerful-looking, juvenile Grey Plover by having a finer bill and a spotted underside, which in Grey Plover is whitish-buff with fine brownish streaking. Note that the tertials are spotted and divided by a narrow, black line at the tip. 15.8. JL.

Widespread overlap between the subspecies occurs in all parts of the range, resulting in many birds showing the characters of either race.

Similar species

Juveniles, in particular resting birds, can be confused with juvenile Grey Plover. However, in flight the latter clearly differs by having black axillaries, a white, barred tail and a white rump. Resting birds may also resemble juvenile Eurasian Dotterel, which has a marked, pale buff supercilium, and a pale horizontal band across the breast.

▲ Adult breeding to non-breeding.
The moult begins on the breeding grounds, but does not commence in earnest until during the autumn migration. After having reached the regular roosting sites (such as the Wadden Sea) the body and wing feathers are renewed before the further journey to the winter quarters is carried on.
    This bird has renewed most if its secondaries and primaries save the P9 and P10 on the left wing and P10 on the right (lower) wing. Note that the remaining worn and faded primaries appear noticeably paler and more brownish compared to the fresh ones. 14.8. JL.

The following general features separate the European Golden Plover from the two rare vagrants, American Golden Plover, *P. dominica*, and Pacific Golden Plover, *P. fulva*. European Golden Plover has a slightly larger and more robust body, and shorter legs, in particular the tibia.

The longest tertial reaches the middle of the tail. The underwing sports white axillaries and white arm coverts. In flight the toes do not project beyond the tip of the tail.

Voice

The call, given by both resting and migrating birds, is a characteristic, soft and plaintive whistle '*toohooh*'. The song, which is performed from a high, butterfly-like display-flight, is a softly melancholic, rhythmic

▼ Adult non-breeding. In the winter plumage the inspiration for its English name becomes evident. The bird has aged on the bill, which is heavier and 0.5 cm longer than that of the juveniles on the facing spread, as well as on the white belly, and the adult tertials, where black tips are edged golden-white. 2.12. KBJ.

Often four primary tips projecting beyond the tertials. (Here one black and three white-edged primaries can be counted).

Tertials and their length. The tertials reach to roughly the middle of the tail.

1–3 primaries extend beyond the tail. Here two, which are counted on the folded, upper wing-tip.

Tip of the tail.

'*pooh-peee-eew*'. From alighting birds, or from flocks, a rolling, repeated '*ploohrrlya*' is also heard.

Habitat

The European Golden Plover can be seen during migration to and from the breeding areas of Scandinavia, Iceland and Russia. Flocks are often seen on grassy fields, meadows, marshes and on ploughed or stubble fields. There they forage in the company of Common Starlings and Northern Lapwings. At other times they rest throughout the day in order to make good use of their large, light-sensitive eyes which enables them to forage during moonlit nights, thus protected from birds of prey.

Outside the breeding season the species seeks out similar habitats for foraging, at the winter quar-

The arrows refer to the features which in all plumages separate the European Golden Plover from the two rare vagrants, American and Pacific Golden Plovers.
    Note that worn, moulted and growing feathers determine the number and length of the visible feathers.

Short legs, particularly the tibia (shin bone), which is often almost hidden by the belly feathers.

ters. Also seen on saline marshland, and on muddy or sandy tidal flats.

Breeding biology

The plaintive call of the European Golden Plover is synonymous with wild, open landscapes. This species does not breed until two years of age, when it seeks out habitats such as moors, raised bogs, tundra and mountainous heaths.

The pair is monogamous and keeps together for life. The nest is established in a shallow hollow in the ground and is lined with parts of plants. The chicks hatch after 30 days, and are independent and able to fly after yet another 30 days. The adult birds leave the breeding areas before the juveniles. The oldest ringed bird lived for 12 years and nine months.

Migration

The spring migration is at its peak in April–May. The birds that stop off at western localities, such as the Wadden Sea, primarily migrate to Iceland and the Nordic mountain areas, whereas birds from the Finnish and Russian populations take a more easterly route. The onset of the breeding season takes place between April and July, depending on the degree of latitude.

The first adults begin their return migration in early July with numbers rising until well into August. The juveniles are the last to leave the breeding grounds, and their numbers at stop-off sites peak in September–October before the migration towards the true winter quarters in western Europe, the Mediterranean countries and north Africa.

The birds of the British population are short-distance migrants and winter in suitable localities close to the breeding areas.

Distribution

The population is estimated at between 630,000 to 860,000 breeding pairs. From a small population in north-eastern Greenland the distribution stretches south and eastwards over Iceland, the north Atlantic Islands, Scotland, Scandinavia, the Baltic and Finland. In northern Russia the species is distributed on the tundra south of the Taymyr Peninsula in Siberia. The largest populations are found on Iceland, Finland, Sweden, Norway, Russia and in the British Isles.

▲ Juvenile in flight.
Note the lack of any signs of the breeding plumage as well as the very fresh flight feathers. 1.9. JL.

▼ First winter.
This small group of resting juveniles has been aged on their dirty-white bellies, retained white-tipped, juvenile coverts, shorter bills and some unmoulted juvenile tertials which have tips divided by a narrow, black line through the tip. The moult from juvenile to first winter is completed by October–November, and from mid winter the juvenile is difficult to tell from the adults. 17.1. KBJ.

PACIFIC GOLDEN PLOVER
PLUVIALIS FULVA

Meaning of the name
'The yellowish-brown rain-caller'
Pluvia, rain. Related to both the autumn migration, which heralds the onset of the rainy period, and the species' behaviour during low pressure conditions, when resting birds huddle together and 'call for rain' with their mournful whistle.

Fulvic, yellowish-brown, referring to the bird's golden-brown upperside.

Jizz
L. 23–26 cm Ws. 60–72 cm.
Significantly smaller, more slender and longer-legged than European Golden Plover. In flight shows greyish-brown axillaries and greyish-white arm coverts that distinguish it from European Golden Plover, which has very light underwings with white axillaries.

In flight. The toes reach some way beyond the tail in flight. On the European Golden Plover, they just about reach the tip of the tail.

Outside the breeding season, may be confused with its close relative, the American Golden Plover, but has a less dainty head shape and heavier breast area. The American Golden Plover has a smaller and rounder, pigeon-like head and slimmer body.

Plumage and identification
Adult breeding, male. Has golden, black and white-speckled upperside, which is separated from the black underside by a wide, white stripe from forehead to the under-tail coverts. When standing, the bird can be most easily distinguished from European Golden Plover by the the jizz: smaller, with slightly longer and more slender body without deep, pigeon-shaped breast and with longer legs. All in all, a smaller and more elegant European Golden Plover.

Female. As the male, but with varying greyish-black front pattern, often with more white on the head and breast.

▼ Adult non-breeding to breeding.
Presumably in transition to full breeding plumage. Khok Kham, Thailand. 28.4. HS.

The arrows indicate characteristics that in all plumages distinguish it from its close relatives, the American Golden Plover and the European Golden Plover. Note that worn, moulting, and growing feathers influence the number and length of visible feathers. Refer to the two other species for more details.

▲ Immature. Similar to adult in winter plumage, but has yellow tinge on head and neck and worn plumage. Birds like this typically stay in the wintering area and first breed when they reach two years of age. Heuksan Do, South Korea. 27.04. AA.

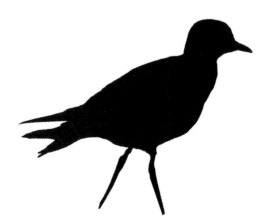

2–3 primary tips behind the tertials. (Here one black-tipped primary and one with a narrow white edge is just visible.)

Tertials and length of tertials. The tertials almost reach the tip of the tail.

0–2 primary tips behind the tail. Here one, which is counted on the folded tip of the upper wing.

Tip of the tail.

Very long, greyish legs, especially the tibia.

▲ Adult breeding, male.
Pacific Golden Plover in breeding plumage and in flight is easy to distinguish from both European Golden Plover and American Golden Plover.

Axillaries on Pacific and American Golden Plovers are greyish-brown with greyish underwings, while the European Golden Plover always has white axillaries in contrast to the remaining parts of the pale underwing as well as the toes that never reach the tail edge.

On the American Golden Plover, the white dividing line stops between the head and neck at the front edge of the wing and the toes reach the tail tip, or are just slightly behind it. 28.4. Khok Kham, Thailand. HS.

Voice

The song is a loud and light, shrill '*dlee-dluuuitt*'. The call is a short '*chu-ittt*', recalling Spotted Redshank, but is longer on the second syllable; '*plir-plooh-eep*'. Heard in several variations and in all cases is harder and shriller than that of European Golden Plover.

Habitat

Breeds on the Siberian tundra from the Urals to the Bering Strait and in western Alaska. Winters mainly in south Asia. Rare vagrant to western Europe, usually in late summer. Often seen in the company of European Golden Plovers on harvested fields, meadows or salt marshes and muddy flats.

▲ Pacific Golden Plover in the foreground to the right, and European Golden Plovers in the background. All breeding to non-breeding.

Despite the heat haze and blurriness, the jizz of the two species is very pronounced and the Pacific Golden Plover stands out clearly as being smaller, more elegant and, particularly, longer-legged than its relatives. Staunings Ø. 1.8. HS.

▼ Adult breeding, male. Very similar in breeding plumage to the northern race of European Golden Plover, *P. a. altifrons*, which has fully marked black face and underside with or without black wedge on the under-tail coverts. However, it can be distinguished from the European Golden Plover by its smaller and slimmer body and longer legs, and from the American Golden Plover by the white line that separates the upperside and underside all the way from forehead to the tip of tail. 7.7. Chukotka, Sibirien. GV.

▼ Adult breeding, female. As with the other species of golden plover females, it has brownish-black markings on the head and is more white on the underside than males, especially on the breast area. Is distinguishable from the European Golden Plover and American Golden Plover by the jizz and the more or less clear white line from forehead to tail. This female has been ringed and flagged at the breeding grounds in Alaska, where American and Pacific Golden Plovers overlap in distribution. 1.6. Yukon. Alaska. GV.

▼ The juvenile Pacific Golden Plover differs from juveniles of European Golden Plover and American Golden Plover by the golden tinge on the head and neck, and especially on the gold-striped crown and the upperside with the wide, golden spots on the feather edges. It is also distinguishable from the European Golden Plover by the slim, more elegant and long-legged jizz. 10.9. Finland. MV.

AMERICAN GOLDEN PLOVER
PLUVIALIS DOMINICA

Meaning of the name
'The rain-caller from San Domingo'
Pluvia, rain. Related to both the autumn migration, which heralds the onset of the rainy period and the species' behaviour during low pressure weather, when resting birds huddle together and 'call for rain' with their mournful whistle.

Dominica, after San Domingo or the Dominican Republic in the Caribbean where the species stops off to and from the winter quarter in South America.

Jizz
L. 24–28 cm. Ws. 65–72 cm.
Slightly smaller than European Golden Plover, but more slim and well-proportioned with longer legs and longer wings. Outside the breeding season it resembles its close relative, Pacific Golden Plover, showing greyish-brown axillaries and greyish underwing in flight. The toes just reach, or project a tiny bit, beyond the edge of the tail.

On standing birds the rounded, pigeon-like shape of the head, the slim body lacking a deep breast, and a long primary projection behind the tail are good pointers towards American Golden Plover.

Refer to the accounts for both European Golden Plover and Pacific Golden Plover for further details.

Plumage and identification
Adult breeding. Refer to the caption on the next page.

Adult non-breeding. Resembles the juvenile, but is basically grey with very little pale spotting on the upperside. Neck, breast and belly are greyish-white and faintly vermiculated. Neither juvenile nor adult has the yellowish tinge to head and neck.

▼ Adult breeding to non-breeding.
Very difficult to tell from Pacific Golden Plover outside the breeding season. The remains of the former wholly black breast and underside are still present on this presumed male. 31.8. Sandy Hook, New Jersey. TH.

The arrows indicate characteristics that in all plumages distinguish it from its close relatives, the Pacific Golden Plover, and the European Golden Plover. Note that wear moult and new feather growth influence the number and length of visible feathers.
Refer to the two other species for similar details.

▲ Juvenile.
Note the long legs, greyish-brown axillaries and underwing as well as few golden fringes on the feathers on the back, and no yellowish tinge to head and neck. 1.12. Israel. LK.

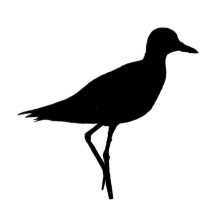

Tertials and length of tertials.
The length of the tertials varies from reaching the middle of the tail to almost reaching the tip of the tail.

Tip of the tail.

2–3 primary tips behind the tail. Here 3, which are counted on the lower, folded wing.

4–5 primary tips behind the tertials. Here 4, counted on the lower, folded wing.

Long, black legs, especially the tibia.

◀ Adult breeding to non-breeding.
Like Pacific Golden Plover, has greyish-brown axillaries and median and lesser coverts contrasting to the rest of the greyish underwing.
　The white 'judge's wig', which ends at the edge of the wing, and the remnants of the former all-black underside (including the undertail), can still be seen below the emerging winter plumage.
31.8. Sandy Hook, New Jersey. TH.

▶ Juvenile American Golden Plover is the most greyish of the juvenile golden plovers, and differs by having a broad, white supercilium, no yellowish tinge to head and neck, and only sparse golden fringes to the scapulars and mantle feathers. 15.10. New York. KK.

Voice

The song is a powerful, short and rapidly repeated '*plee-eeck*' with a dry cracking on the last syllable. On landing a springy, rolling '*veet-veet-dooh-eeu-veet-veet*'. The alarm call is a strong '*dlooh-kee-kee*' and the call is a shorter or longer, soft '*dlooh-eet*'.

Habitat

Often seen in company of European Golden Plovers on harvested fields, grasslands or on costal meadows or muddy flats.

Distribution

Rare vagrant from North America to western Europe, most often in late summer. Breeds across the North American continent from western Alaska through Canada to the Baffin Island.
　Winters in South America, primarily in Brazil, Paraguay and Argentina.

▼ Adult breeding, male to the left and female to the right.
The male in breeding plumage differs from European Golden Plover as well as American Golden Plover by its all-black underside and the white 'judges wig' which ends at the carpal joint.
　The other two species have a white demarcation line between the upperside and underside from forehead to tail. The female is, as in the other species, more brownish and diffusely marked than the male. Alaska. 15.6. KK.

GREY PLOVER
PLUVIALIS SQUATAROLA

Meaning of the name
Pluvia, rain. Related to both the autumn migration, which heralds the onset of the rainy period, and the species' behaviour during a low pressure period, when resting birds huddle together and 'call for rain' with their mournful whistle.

Squatarola, a Venetian name for a certain kind of plover.

Jizz
L.27–31 cm. Ws. 71–83 cm.
Medium-sized wader with a thick and heavy bill, and the largest of the Pluvialis plovers. Most often seen, as shown in the silhouette, in a slightly hunched-up position, and usually keeping a good distance from conspecifics and other waders. This

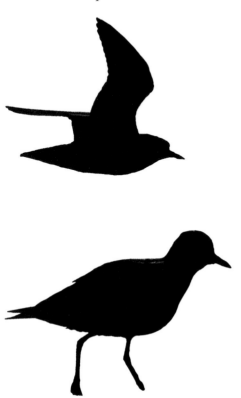

'lone wolf' tendency makes it easy to spot exclusively by shape and behaviour.

In flight. The flight is powerful. Often flies alone or in small groups, and is distinguishable from all other species of wader by having black axillaries in all plumages. It is also the only Pluvialis plover and plover in general with a hind toe.

Plumage and identification
Adult breeding, male. Unmistakeable in its smart, black-and-white dress with the upperside coarsely speckled in black and white, white or black-flecked crown and nape and white sides of the breast. The face, breast, belly and legs are black. Behind the legs, the vent and under-tail coverts are white.

In flight. The upperside shows a prominent, white wing-bar (the most pronounced among the *Pluvialis* plovers). The rump is white, and the tail white with narrow black bars. The underwing is white with diagnostic black axillaries.

Female. As the male, but as is typical for *Pluvialis* plovers, shows less extensive blackish-brown markings, often admixed with pale feathers.

Adult non-breeding. Refer to caption.

Juvenile. Undergoes a variable moult from autumn to later in the spring. The upperside is coarsely spangled in blackish-brown and buffish-white, not golden as in juvenile European Golden Plover. The head, neck, breast and belly are streaked and have a faint, yellowish-buff tinge. The vent is white and the legs are greyish-black.

Adult non-breeding to breeding. Moulting the head and body feathers to acquire the complete breeding plumage. The sexes are best told apart on the breeding ground, but the blackish-brown underside of the female is a good indicator. 16.5. HS.

▲ Adult breeding, male. 15.8. JL.
▼ Adult breeding, male, with a few retained, worn (brownish and faded) tertials, scapulars and coverts. The male in complete breeding dress is one of the most striking of the Arctic waders. Both sexes are surprisingly well camouflaged in adaptation to their habitat, which is the part of the tundra where aler greyish lichens predominate.17.6. Norway. DP.

▲ Adult breeding, female.
Still in breeding plumage as is the flying male to the left.
The moult to non-breeding commences in August and lasts until the late autumn, in some individuals until December or until the following spring. 15.8. JL.

▲ Juvenile. The diagnostic black armpit makes an eye catching contrast to the white underwing, and is a safe identification feature in all age groups and plumages. Juvenile birds are heavily streaked with a yellowish-buff wash on head, breast and belly.
The upperside is greyish-brown and coarsely speckled with a prevalence of pale buffish edges and some golden ones. 21.9. JL.

Voice

The flight call is most often heard during the migration. It is a soft, mournful whistle, dropping in pitch in the middle; '*ploohooh-vee-ee*', quite different to the call of the European Golden Plover, which is a shorter '*toohooh*'.

The song is delivered in flight with slow, stiff wing-beats and is a drawn-out '*plu-ee-eew*'.

The juvenile differs from adult and first winter by having a distinctly spotted upperside in blackish-brown with white and (some) yellowish spots, as well as by the underside which has a yellowish-buff tinge. Told from the other juvenile *Pluvialis* plovers by its more powerful body, the heavier bill and the pure white vent. 21.9. JL.

Habitat

Outside the breeding season it appears scattered, turning up singly or in small, loose groups along coasts with muddy flats, shallow sandy lagoons and on beaches with banks of seaweed; occasionally also in similar inland habitats. Larger flocks occur in the regular winter quarters. Some non-breeding birds remain in the winter quarters throughout the summer, for example in the Wadden Sea. Individual birds defend a feeding territory against competitors.

Breeding biology

Breeds when two or three years old. The pair is monogamous and sticks together for years, often on or close to the same breeding place with a distance of several hundred metres to the nearest neighbouring pair.

The nest usually contains four eggs, and the clutch is laid in a shallow depression lined with moss, lichen or pebbles. The eggs are incubated for *c*. 26 days by both parents and after hatching the adults lead the young for up to three weeks. Thereafter the brood fledge and become independent. The oldest ringed bird lived for 25 years and seven months.

Migration

The spring migration through Europe towards the Arctic tundra begins in April. Regular roosting sites in, for example, the Wadden Sea are used for the birds to store fat reserves before undertaking the last leg of the journey. The breeding grounds are reached from late May to early July. Departure from the breeding areas takes place from late July to September. Firstly the adult birds leave, followed by the juveniles from August to September.

European birds from the north-western and north Russian population migrate along the coasts of western Europe towards western Africa until the end of November. In mild winters smaller numbers remain in western Europe.

Eastern populations migrate along the east Mediterranean flyway to eastern Africa and South Africa.

Distribution

The European population, which solely breeds west of the Ural river, numbers between 5,000 and 10,000 pairs. The species is seen on migration in spring, late summer and autumn. Excluding Greenland, it is an almost circumpolar breeder on the tundra along the Arctic Ocean. Until recently monotypic, but has now been divided into three subspecies primarily due to geographical variation. The nominate form breeds from the most northerly parts of north-western Russia and east across the continent through Asia and in western and northern Alaska, with winter quarters from western Europe and Africa to southern Asia and Australia.

Subspecies *tomkovichi* breeds on the Wrangel Island off north-eastern Siberia with presumed wintering areas along the east Asian coasts.

Subspecies *cynosurae* breeds along the coasts and on the islands in Arctic Canada with winter quarters along the coasts of North and South America.

▶ Adult non-breeding, *cynosurae*. Adult differs from juveniles in having a more smudged grey appearance, without the distinctly white spotted upperside of the juvenile. 22.12. California. DP.

▼ Adult non-breeding.
Below the adult tertials with narrow, pale edges, the tips and edges of emerging, fresh primaries are just visible, while a couple of old, brownish and worn ones stick out behind the tail. Note the difference of the plumage and the individual feathers between this bird and the juvenile to the left. 13.11. KBJ.

Scolopacidae is the largest wader family, comprising 16 genera, including the *Calidris* sandpipers, the snipes, the *Tringa* sandpipers, the curlews and the phalaropes.

These waders are found throughout the entire globe except the Antarctic. Most species breed in the Northern Hemisphere and are long-distance migrants.

This is a highly varied family, including birds of odd shapes and sizes as well as some with showy colours. The family includes the Least Sandpiper, smallest wader in the world, the peculiar and rare Spoon-billed Sandpiper of east Asia, the unique and colourful Ruff, and the largest wader in Europe, the Eurasian Curlew with its long, curved bill.

In the following chapter the species which breed and migrate through Europe are described, as well as the most often observed American and Asian vagrants, together with some possible future vagrants.

Three representative members of Scolopacidae on a high tide roosting place in the Wadden Sea.
All are adults in breeding plumage on their way to the Nordic and Arctic breeding places.
Bar-tailed Godwit with copper-red males and slightly larger, pale females, Red Knots, mostly in breeding plumage, near the top of the frame, and a few Dunlins. 4.5. LG.

The *Calidris* sandpipers number 24 species, all breeding in the northern hemisphere.

Six species breed within Europe, while an additional three are regular migrants. Only three of the remaining 15 North American and Asian species have not yet been recorded in Europe.

The *Calidris* sandpipers make up the second largest wader genus after the banded plovers, but when it comes to challenges for the birdwatcher, the *Calidris* species are second to none.

Their size varies from the smallest of waders, the seven stints or peeps, which are all very similar, particularly in winter plumage, through the Common Starling-sized Dunlin, which is the world's most common *Calidris* with an estimated population of at least 5 million individuals, to the robust Great Knot which breeds in Siberia.

Throughout the year, the *Calidris* species go through various stages of moult approximately every third month, displaying an assortment of intergrades between from greyish-white winter plumage to spangled, often colourful breeding plumage.

During the spring and autumn migration, the *Calidris* species appear in almost all their plumages on their way towards the Nordic and Arctic breeding areas. Winter quarters are primarily in the Southern Hemisphere. They thus provide constant experiences, challenges and frustrations for birdwatchers in general.

The *Calidris*-like Ruddy Turnstone is one of two turnstone species within the genus *Arenaria,* and the family Scolopacidae. Its account in this book is placed between the European *Calidris* species and the rare American and Asian *Calidris* vagrants.

◄ Post juvenile moulting to first winter.
A young Dunlin is a good example of the complexity of plumages that the many very similar *Calidris* sandpipers exhibit, and of the challenges facing the birdwatcher when tackling species identification and aging.

This bird shows typical juvenile *Calidris* traits, namely a fresh plumage with rusty-red, buff or whitish feather fringes, as well as tertials with narrow, coloured edges. A detail which supports the identification of this bird as a juvenile is the moderate wear of the pale buff tips of the median coverts and the bottom row of smaller coverts.

That the bird is entering its first winter plumage is revealed by the row of new, grey scapulars with black shaft streaks.

Identification to Dunlin is reached through the combination of overall impression of size, colour of legs, the patterns of the plumage and the crucial black, ventral spots which gives away the species. These scattered and often diluted spots are suggestive of the full black belly patch adorning the adults in breeding plumage. 6.9. LG.

DUNLIN
CALIDRIS ALPINA

Meaning of the name
'The sandpiper from the high mountains'
Calidris, from Greek *kalidris* or *skalidris*, name for a greyish bird along the water, mentioned by Aristotle.

Alpina, from high mountains. Refers to the breeding habitat on high-altitude plateaus.

The fact that it is the most numerous of its genus, and the most common in Europe, makes it the obvious species of reference for identification of *Calidris* species specifically, but also for identification of smaller waders in general.

▲ Adult breeding. *C. a. alpina*, on return migration. In flight the Dunlin is the size and shape of a Common Starling, although it has longer wings and bill. In breeding plumage it is unmistakable among the *Calidris* species and other smaller waders, as the only species with a large black belly patch. Large flocks suddenly flash white and seemingly almost disappear when all individuals turn at once. 2.8. PN.

▼ The species is divided into 10 subspecies, which show considerable plumage variation. In particular the intensity and combination of rusty-red and black varies. Depicted here is a presumed *C. a. alpina* on its way eastwards to Russia, perhaps Siberia. The bird has been sexed as a male due to its grey nape, which makes a contrast between the crown and the mantle. 26.5. Gotland. Sweden. DP.

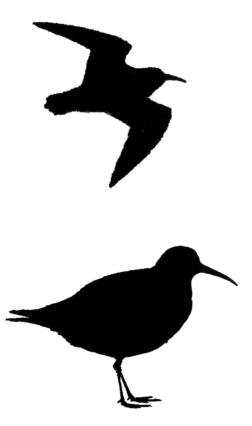

Jizz

L. 16-22 cm. Ws. 33-40 cm.

Common Starling-sized with rounded and compact body, short neck and medium-long bill which varies in length, and with the outer third curved downwards. Medium-long, black legs.

The only small wader with an all-black belly patch in the breeding season.

Standing birds often appear hunched-up and short necked.

Often forages in flocks with energetic and rapid movements, either by snatching food items from the surface, or probing at the ground with fast, sewing machine-like movements. Often wades belly-deep in water.

Typically seen in flocks during migration, and outside the breeding season often together with other smaller waders.

In flight. Has fast, whirring and direct flight, often with sudden twists and 180-degree turns, when the white underside will flash conspicuously. The toes reach to the tip of the tail, or slightly beyond it.

Similar species

In winter, juvenile and transitional plumages the Dunlin can be confused with three European *Calidris* sandpipers of similar size: Broad-billed Sandpiper, Sanderling and Curlew Sandpiper. However, the shape of bill, head pattern, wing-bar and tail markings separate the four species.

Also compare to rare *Calidris* sandpipers, pages 226-251.

Plumage and identification

As in most *Calidris* species, the plumage changes successively throughout the year, and furthermore varies in the details between the three subspecies most commonly seen in Europe.

Adult breeding, male. Has streaked, rusty-brown crown and more or less reddish-brown to

▲ Adult breeding. *C. a. schinzii.*
Compared with the nominate, it is paler with rufous to yellowish-edged scapulars with greyish-white tips, as can still just be seen on this rather worn plumage. 23.7. Outer Hebrides, Scotland. EFH.

▼ Adult breeding *C. a. arctica.*
This form breeds in north-eastern Greenland and on Svalbard, and is the palest and smallest of the 10 subspecies.
This subspecies migrates across Iceland and along the British Isles and the channel area towards the winter quarters in western Africa. 25.6. Svalbard. HS.

▲ Adult breeding. *C. a. alpina.*
In its northern breeding distribution, the species lives on mountainous heaths and boggy tundra.
This bird is a female, determined on the streaked, greyish-brown nape and the long bill, which is on average 3-5 mm longer than that of the male. 29.6. Varanger, Norway. HS.

▼▶ The four eggs approach walnut size, and are surprisingly large for such a small bird. This large size, together with the rather long incubation period of *c.* 22 days, is necessary to ensure that the chicks are fully developed and mostly self-sufficient after hatching. 13.6. Varanger, Norway. HS.

minent greyish-black spots; at times so well marked that it comes across as an adult-like toned-down belly patch.

Subspecies

The Dunlin is currently divided into 10 acknowledged subspecies of which three occur in Europe. One is the nominate *C.a. alpina*, which stops off in large numbers on European coasts on both spring and autumn migration, on its way to and from the breeding areas in northern Scandinavia and further east through northern Russia to Taymyr in Siberia. The second one is *C.a. schinzii*, which breeds in south-eastern Greenland, Iceland, the Faroe Islands, the northern British Isles, in southern Scandinavia, and in the countries along the the southern coast of the Baltic Sea. The third is *C. a. arctica*, which breeds in north-eastern Greenland and on Svalbard with its migration route across Iceland, the British Isles and western France.

Furthermore, from time to time very long-billed individuals are recorded, presumably vagrants of other subspecies from eastern Siberia or North America.

golden-brown mantle and scapulars with black feather centres and pale edges. The upper breast is white and finely streaked or spotted with black, contrasting to the back belly patch which extends to the black legs. The rest of the underside is white.

In flight. Has reddish-brown scapulars set against the greyish wings, with a broad, white wingbar, white rump and grey tail feathers with a dark central longitudinal bar. The underwing is white.

Female. Best distinguished from the male on the breeding ground, but is slightly larger with bill 3–5 mm longer (on average). In the Greenland and the European subspecies, females also show a brownish-buff, finely streaked nape, lacking in contrast between crown and mantle, in contrast to the geryer nape of males.

Adult non-breeding. Gives a greyish/whitish impression, with head, sides of breast and upperside uniformly greyish-brown. The upper breast may show fine streaking on the sides.

Juvenile. Fresh juvenile plumage is somewhat reminiscent of a slightly paler and dingy version of the adult breeding plumage, but differs in having quite fresh and tidily placed feathers with a white 'V' on the back, formed by the upper scapulars which, like the remaining scapulars, have pale edges. Also has reddish-buff fringes to coverts and tertials.

The upper breast is finely streaked. The belly and the flanks down to the legs have more or less pro-

The song is a dry whirring trill '*trrryytrryy trryytrryy*', either performed in flapping song flight or on the ground near a female, as the singing male raises his wings high over his back.

Breeding biology

The European birds arrive to their breeding areas from late March and are already incubating from late April. Arctic birds reach their breeding areas as late as the beginning of July.

Dunlins first breed when one or two years old, and the pair is monogamous. One clutch of four eggs is laid in a nest hidden in a tussock, preferably with a roof of overhanging tall grass, and lined with grass and small leaves. The parents share incubation (lasting for 20–24 days) and rearing the young, although the female departs from the territory before the male.

The young are able to fly after 24 days. The oldest recorded bird was almost 29 years old.

Habitat

Dunlins are commonly seen along the coasts in the migration season, where they rest and forage in small flocks on both stony beaches with seaweed as well as on sandy flats and muddy lagoons.

The Wadden Sea and the Wash in eastern England are of international importance as roosting, moulting and wintering areas for the species, and flocks of up to 100,000 birds can be seen at either site in the migration season. The species can also be seen visiting inland habitats such as meadows and farmland, often in company with European Golden Plovers and Northern Lapwings.

Eastward-migrating birds that use the migration flyway over the Mediterranean and the Black Sea towards the Russian and Siberian breeding grounds are also seen at lakes and inland deltas. Dunlins breed on open grassy and heather moorland, and tundra.

Voice

The call is a trilled '*trrrryyy*'.

From foraging birds rapid, crunching contact calls are heard, such as '*brie-bra-bry*', and from roosting birds a rush of contact calls sounding like a heavy shower.

▲ Juvenile. Told from other similar sized juvenile *Calidris* species by a combination of pale rump with a dark central longitudinal bar, the shape of the bill and particularly by the buffish upper breast with streakings that turns into spots or a black smudge on the sides of the lower flanks. 16.8. JL.

▼ Juvenile. In fresh plumage resembles an adult in breeding plumage, but differs overall by the quite fresh and unworn feathers with white fringes to the scapulars. This creates a white 'V' on the back, similar to that of Little Stint. It also has buff fringes on the coverts and whitish to buffish edges to the tertials. The unworn primaries are tipped white. In other words, a typical juvenile plumage. The clinching character, however, are the greyish-black spots that on this individual almost converge to form a belly patch. 9.8. LG.

Migration

The spring migration through Europe takes place from the middle of March to early June, culminating in the middle of May when hundreds of thousands of birds pass. Several hundred thousand birds roost in the Wadden Sea in the spring where they moult into full breeding plumage and accumulate extra fat reserves.

The passage consists of two waves. The first is of birds that have primarily wintered in western Europe, and which breed in Iceland and in northern Scandinavia. Later, birds from the west African winter quarters arrive and they presumably belong to the Russian population.

However, the majority of the birds that winter in Africa are migrating in a broad front up across eastern Europe towards Siberia.

As early as in July the first adult birds are seen on the return migration. Scandinavian as well as Russian and Siberian birds migrate in great numbers via the Baltic Sea towards western Europe via the Wadden Sea. Here, numbers of adults and juveniles peak in August and September and Dunlins can be seen in vast flocks of up to 100,000 birds.

At the end of October most birds have migrated to winter quarters in the British Isles, the western coast of France, the Mediterranean, and primarily Banc d'Arguin in Mali on the north-west coast of Africa. The majority of the Siberian birds winter from the Arabian Peninsula and eastwards to China and Japan, and in south-east Asia.

Distribution

The European breeding population amounts to between 419,000 and 547,000 pairs. The largest populations are found in Iceland where the subspecies *C. a. schinzii* breeds, as well as in Russia, Norway, Sweden and the British Isles. Beyond northern Europe, the species has a circumpolar distribution in the low Arctic part of the Northern Hemisphere.

▲ First winter. The juvenile body feathers are moulted during the late summer and autumn, and moult is completed in the winter quarters through December. Age here is determined by the line of retained juvenile upper scapulars with rusty-buff edges, as well as by the dark-tipped tertials and greater coverts. 19.1. KBJ.

▼ Adult non-breeding. Adults start their moult in the breeding area, suspend it during migration, and complete it on the migration staging posts in the Wadden Sea. Most are in winter plumage from October.
Age determined by the uniform feathers of the upperside with no obvious differences in coloration, nor any remains of juvenile feathering. KBJ. 20.1.

▲ Non-breeding.
Dunlin in winter plumage can be confused with Curlew Sandpiper in juvenile or adult non-breeding plumage. Both species show a generally greyish plumage, and they both have a rather long, down-curved bill. However, Curlew Sandpiper has a pure white rump in these plumages. 17.2. BLC.

▼ Dunlins in almost all variations of plumage can be seen in the Wadden Sea in August and September. The numbers of autumn-migrating Dunlins are at their peak at this time. Both adult and juvenile birds are resting to moult their body feathers before the continued migration to the winter quarters in south-western Europe and in west Africa at Banc d'Arguin in Mauritania.

The area, which is a National Park and UNESCO World Heritage Site, is the most important wintering area for birds on the east Atlantic flyway/migration route, with up to 2 million wintering north European and Arctic waders. The picture below is part of a flock of 5,000 Dunlins at their high-tide roost in the Danish part of the Wadden Sea. 28.9. LG.

BROAD-BILLED SANDPIPER
CALIDRIS FALCINELLUS

Meaning of the name

'The hook-billed sandpiper'

Calidris, from Greek *kalidris* or *skalidris*, name for a greyish bird along the water, mentioned by Aristotle.

Falcinellus, diminutive of falx, sickle, referring to the bird's slightly sickle-shaped bill.

Jizz

L. 16-18 cm. Ws. 34-37 cm.

Resembles Dunlin, but is smaller, shorter-legged and appears more compact and reserved, particularly when roosting or seeking food in a hunched-up position with retracted neck.

However, the broad bill with the faintly curved tip, as well as the white crown stripes bordering the narrow, dark crown, sets it apart from all other smaller *Calidris* species.

Most often seen singly or a few together, usually in the company of other *Calidris* species.

If alarmed it may press itself against the ground like a snipe.

Plumage and identification

Adult breeding. Has a broad bill with slightly curved tip, black lores and prominent whitish supercilium in contrast to a dark crown with pale split supercilium, which borders the crown and extends to the back of the head.

The head pattern is diagnostic and present in all plumages, if a bit more diffuse in the winter plumage. The upperside is blackish-brown with broad, pale buff edges that have a rusty-red tinge on the mantle, scapulars and tertials.

The sides of the breast have a ruddy wash and dark, triangular spots. The rest of the underside is white and the legs greyish-green or greyish-black. In the course of the breeding season the pale feather tips are gradually worn off, rendering the bird very dark.

Adult non-breeding. Mainly pale. Greyish-brown upperside with narrow, dark shaft streaks and streaked neck against the pale underside.

Juvenile. Very similar to adult breeding, but sports a fresh plumage with neatly arranged feathers.

Subspecies

Two very similar subspecies. Refer to distribution.

Adult breeding.
Note the characteristic head pattern, the bill and the often more crouching stance, which sets it apart in a mixed flock of *Calidris* sandpipers. 29.5. Sweden. DP.

Voice

The flight call is a short, almost soft trill, '*plllrett*', at times dual, '*plllrett-plrt*'.

The song, delivered from song flight, resembles a rapid, pulsating repetition of the call '*dlllrruee-dll-rruee-dlllrruii*', often with a whirring, Greenfinch-like trill inserted.

◀ Juvenile with golden and rufous edges on mantle feathers and scapulars. Has a white wing-bar like the other smaller *Calidris* species. Furthermore has a white 'V' on the back as in juvenile Dunlin and Little Stint, but is easily distinguished by the diagnostic head markings and the slightly hooked bill. 15.9. Israel. LK.

▼ Juvenile. The English name is appropriate when the bird is seen from the front. Note the characteristic head markings, which separate it from the other *Calidris* species. 18.8. HS.

▼ Juvenile is told from adult by its fresh and neatly arranged feather tracts with broad, brownish-buff and white frayed edges. Note the lower mandible, which on both adult and juvenile is tinged yellowish on the inner part, as well as the leg colour, which is greenish-yellow on the juvenile. 20.8. HS.

Habitat

When migrating, the species is seen occasionally along muddy lagoon shores and tidal flats, as well as shores of larger lakes. The breeding habitat is sparsely vegetated upland moor and tundra with small ponds.

Breeding biology

The pair is monogamous, and may breed in small, loose colonies. The male sings in display flight over the territory, and defends it rigorously against rivals.

The single clutch of usually four eggs are laid in a lined nest, often on the top of a tussock of moss. Incubation lasts for a good 22 days, and both sexes share in the parental care, but the female leaves the family before the young fledge.

The oldest ringed bird lived for 10 years.

On migration in western Europe and Scandinavia, the Broad-billed Sandpiper is mostly seen singly or in small groups, usually in company with other *Calidris* species and other small waders.

In this small spring flock of Common Ringed Plovers, Dunlins, one Curlew Sandpiper and two Broad-billed Sandpipers, all birds are in breeding plumage.

Can you spot the Curlew Sandpiper and the two Broad-billed Sandpipers? 13.5. DP.

Migration

Scarce spring migrant in western Europe from the middle of April to early June.

Adult birds are already on return migration from early July, and the juveniles appear from the end of the month.

The migration from the north European breeding areas passes south-east on a broad front, primarily overland and partly along the coasts. The species migrates alone or a few together, although larger flocks can be recorded, for example during the autumn migration along the island of Öland in Sweden, as well as on well-known migration staging posts such as Sivash in southern Ukraine near the Black Sea. The north-western populations winter in east Africa, at the Red Sea and in Arabia, as well as in western India and Sri Lanka.

Birds of north-eastern origin spend the winter in southern Asia to Australia.

Distribution

Between 29,700 and 44,100 pairs breed in Europe, in Scandinavia and Russia, with Finland accounting for 82% of the population. The nominate subspecies breeds in northern Fennoscandia, and in north-western Russia.

The subspecies *C. f. sibirica* is distributed in northern and north-eastern Siberia.

▲ Adult non-breeding in flight. Overall greyish-brown and is very similar to Dunlin in winter plumage, but can be distinguished by the dark lores and the diagnostic crown markings, which are still discernible in the winter season.

However, Broad-billed Sandpiper in non-breeding plumage is rarely encountered in Europe as by then it has long since departed for Africa, Arabia or southern Asia. The bird seen here is aged by the completely fresh remiges, which are mostly moulted after arrival to the winter quarters, and by the absence of buff or rusty tinged juvenile feathers. 17.1. Oman. HJE.

▶ Adult breeding to non-breeding.
When migrating, the moult of the body feathers begins on the migration staging posts, then is suspended during the continued journey to be completed in the winter quarters.

The plumage shown here mainly consists of new, grey, non-breeding scapulars, head and mantle feathers.

Furthermore, note a single brownish, worn and faded greater covert as well as equally worn flight feathers. 24.8. Oman. HJE.

CURLEW SANDPIPER
CALIDRIS FERRUGINEA

Meaning of the name
'The rusty-red sandpiper'
Calidris, from Greek *kalidris* or *skalidris*, name for a greyish bird along the water, mentioned by Aristotle.

Ferruginea, rust-coloured, *ferrum*, iron, *ferruginis*, rust of iron.

Jizz
L. 18–23 cm. Ws. 38–41 cm.
Very similar to Dunlin in shape, but is more elegantly built with a well proportioned, elongated body. It has a longer neck, wings and legs than Dunlin, and most importantly a longer and more curved bill, particularly on the outer third. Often seeks food in deeper water, and can from time to time be seen swimming.

In flight. The Curlew Sandpiper is told from the other similar-sized European *Calidris* species by the white rump of juveniles and adults in winter plumage. Furthermore the toes project a little beyond the tail.

Adult in breeding plumage has coarsely black-spotted rump.

Plumage and identification
Adult breeding. Unmistakable in its rusty-red plumage. Has white around the base of the bill and often around the eye. The crown is streaked black and golden, while the rest of the head, neck and breast is rusty-red. The vent is white-spotted and the legs are black.

The upperside is speckled in greyish-black with greyish-buff fringes and orange-red spots.

The underwings are pure white set against the red body. On the breeding ground the female can, in some instances, be told from the male by a slightly longer bill and weaker patterns on the breast with variable barring.

Non-breeding. Has a streaked neck and crown and a prominent supercilium. The rest of the underside is white.

The upperside is greyish to greyish-brown with narrow, dark shaft-streaks and in flight it shows a pure white rump, contrasting with the grey tail.

Juvenile. Has a greyish-buff scaly upperside with pale fringes. The neck is finely streaked, and nape and head with pale supercilium. The breast is buff, often orange-tinged and the belly is white. Pure white rump in flight.

Similar species
Adult in breeding plumage may be confused with the larger and robust Red Knot. In non-breeding and juvenile plumage, could be confused with adult and juvenile Dunlin.

▼ Adult non-breeding to breeding.
The emerging rusty-red breeding plumage can just be seen on the breast, and also the scapulars with a few emerging rusty-red, black and grey feathers appearing. 20.5. Greece. DP.

▲ Left bird is breeding to non-breeding, but still with characteristic white rump with coarse black spots, and a rusty-red belly. To the right is a juvenile with characteristic scaly upperside and bright, pure white rump, which distinguishes the species from all other juvenile *Calidris* species.

Juvenile Ruff can seem similar, but in that species only the sides of the rump are white, sometimes forming a white horseshoe on tail and rump. 12.8. JL.

▼ Adult breeding, still in full breeding plumage, but with very worn coverts and scapulars.

The moult of the worn body feathers begins on the migration staging posts, while the flight feathers are not shed until the migration has ended and the winter quarters have been reached. 11.7. Sweden. DP.

Voice

The song is heard both from the ground when the male is accompanying the female, and from a song flight performed high as well as low over the territory. It is introduced by some short, hard pricking '*pripp*' notes, quickly followed by an accelerating trill, which at times ends with a shrill, ascending whistle.

The flight call is a short, trilling '*choohrrip*' in two parts.

Habitat

Does not breed in Europe, but is seen on migration, which takes place over land. Birds travel alone, in small groups or in mixed flocks of other waders, most often Dunlins and Ruffs.

Roosts and winters on and along muddy, sandy or gravelly shores, flats and salinas, both along coasts and rivers as well as at larger lakes.

Breeding biology

Breeds on Arctic tundra with damp hollows and small ponds. The male can be seen pursuing the female in a low or high courtship flight over the territory. As soon as the eggs have been laid, the male leaves the territory. The single clutch of usually four eggs are incubated solely by the female for *c.* 20 days. After hatching the chicks are led to an area with higher vegetation and rich insect life, often in loose association with other females that jointly defend the offspring against predators.

The breeding success is linked to the population of lemmings, which generally peak every fourth year. When the lemming numbers are highest, the

▲ Juvenile.
The neatly scaled upperside is greyish-brown with a brassy tinge, and the fringes are pale grey. The faintly streaked orange-buff breast and the long, slightly curved bill easily distinguish it from other juvenile *Calidris* species. 21.8. NLJ.

▼ Adult breeding.
The species is typically seen in muddy coastal inlets and at larger lakes. The flock seen here in Denmark has presumably arrived from Siberia via the White Sea, the Gulf of Bothnia and further down through the Baltic Sea. 17.7. ISA.

survival of the chicks is high, but in years with very few lemmings, predators like skuas and foxes take many more young waders instead.

The young are able to fly after about 16 days.

The oldest ringed bird lived for 19 years and eight months.

Migration

The northward spring migration primarily passes overland through north Africa and eastern Europe. The species is thus seen in small numbers in western Europe from late April to early June, and again in larger numbers from the middle of July when the adults begin the return migration.

The migration of juveniles culminates from the middle of August to early September, and the last juveniles leave the continent through October. A small part of the population – fewer than 2,000 individuals – winters on the Iberian Peninsula.

The species is a long-distance migrant, and the Siberian birds, which are seen on autumn migration in Europe, primarily follow three important migration routes towards the winter quarters in Africa.

One route passes along the coasts of the White Sea overland to the Baltic Sea and further on along the west European coasts to western Africa.

Another route passes to western Africa passes over eastern Europe and the Black Sea.

The third route passes over the Black Sea, the Caspian Sea and the Middle East to eastern and southern Africa.

Birds from eastern Siberia winter from India through southern Asia to Australia.

Distribution

Breeds on the tundra or along the coasts of the Arctic Ocean in northern Siberia.

▲ Breeding to non-breeding.
Has worn tertials and very worn wing coverts with exposed feather shafts, but winter plumage is developing, with fresh, grey feathers on the mantle and individual new large, grey scapulars. The remains of the brick-red breeding plumage can be seen on the underside. 16.8. JL.

▼ First winter.
Told from adult non-breeding, which has grey upperside feathers with narrow, black shaft streaks, by the row of retained, brownish-beige scapulars and the brownish-beige, white-fringed wing coverts as well as the fresh, pale-edged tertials. 12.11. Oman. ISA.

RED KNOT
CALIDRIS CANUTUS

Meaning of the name
'King Canute's sandpiper'
Calidris, from Greek *kalidris* or *skalidris*, name for a greyish bird along the water, mentioned by Aristotle.

Canutus, named after King Canute (Canute the Great, 995-1035), who was the king of Denmark, Norway and England and according to the legend liked to eat Red Knots stuffed with white bread soaked in milk.

Jizz
L. 23–25 cm. Ws. 45–54 cm.
A robust *Calidris* with a somewhat pigeon-shaped body, a rather short neck and a powerful, short and straight bill. Relative to the size of the body, the medium-short legs gives the bird a thickset appearance.

The species is one of the largest *Calidris* sandpipers in Europe and it thus clearly stands out in mixed, migrating flocks of smaller *Calidris* Sandpipers. Forages in typical *Calidris* manner, with a nodding gait. Is rusty-red overall in breeding plumage, in winter plumage light grey.

In flight. Has a compact, elongated body with long, slender wings and quite measured wingbeats for a *Calidris* species.

Similar species
Adult breeding. May be confused with the brick-red Curlew Sandpiper, which is more slender and has a long, thin and curved bill, and also with the Bar-tailed Godwit, which has long legs and a long,

▲ Adult breeding, *C. c. canutus*, on its way towards Siberia after a good three weeks of over-eating in the Wadden Sea, where its weight has been doubled in the form of fat, which constitutes the fuel for the long, last leg of the direct, nonstop migration of c. 4,500 kilometres. Before the stop in the Wadden Sea the bird has migrated for an equal distance from the winter quarters in west Africa. The total distance of more than 9,000 kilometres in two 'legs' is one of the longest wader migrations. The brick-red head and underside in contrast to the greyish upperside with a short wing-bar and barred uppertail coverts make the species easy to identify at a long distance.

The only possibility of confusion on the migration is Bar-tailed Godwit, which has a long bill and long legs.
2.6. Öland, Sweden. JL.

slender bill. Red Knots and Bar-tailed Godwits are often seen together in smaller groups, as well as in larger, mixed flocks at the migration staging posts such as in the Wadden Sea.

Plumage and identification

Adult breeding. Mantle, scapulars and back are speckled in black and brick-red, often with a few still unmoulted, grey winter feathers. Head, neck and breast are brick-red. The lower part of the belly and the vent are white with scattered red feathers. The legs are greyish-green.

In flight. Has a black and red-speckled back in contrast to greyish wings, a narrow white wing-bar and a grey tail with barred tail coverts, but (unlike most *Calidris* species) without a dark central bar.

Adult non-breeding. Has a dark lore and a broad, pale supercilium against a pale grey crown. The rest of the upperside is greyish-brown with dark, thin shaft streaks. The throat, neck and breast are finely streaked grey, while the belly and flanks are whitish-buff with scattered, grey vermiculation.

Juvenile. Has a beautiful steel-grey upperside with large, white-fringed feathers with black subterminal bars. Neck, breast and flanks are finely streaked grey on a whitish-buff background, which in some individuals may have a faint, rusty tinge.

Subspecies

The species is made up of six subspecies, of which only two are seen in Europe. One is the nominate, the other is *C. c. islandica*. The former breeds in the central parts of northern Siberia and Arctic Canada, the latter in north-east Greenland. The remaining subspecies have a Siberian and North American distribution. Separation of the individual subspecies is difficult and is best done by measurements taken in the hand.

Voice

The call is a short, slightly hoarse '*whet-whet*'.

The song consists of bubbling and plaintive sounds: '*duui-ik-duui-ik-duuuieee-duuuieee*'.

The song is performed from lengthy soaring flights interspersed with steep rises and downward gliding on stiff wings and spread tail.

▼ The juvenile is easily told from the adult by its coarsely scaled upperside with buff and whitish fringes. Some individuals have a pure whitish-buff belly, others, such as this bird, have a tinge of the brick-red colour of the breeding plumage in the face and on the breast. 22.8. BLC.

▲ The juvenile in flight is told from other juvenile waders by a combination of the barred uppertail coverts, the short, powerful bill and the robust body. 17.8. JL.

◄ Adult breeding, *C. c. islandica*.
In spite of the name the species does not breed in Iceland, but a part of the Greenland and Canadian breeding population does stop off during return migration on Iceland; an important pit-stop before the last, direct migration to north-eastern Greenland or to Canada and north-east Greenland with a migration route straight over the inland ice. Iceland. 6.6. DP.

Habitat

Seen on migration in spring and autumn in small groups, often together with other waders on sandy, muddy and gravelly coastal stretches as well as in larger flocks on regular migration staging posts, for instance at the Wadden Sea, where small numbers also overwinter in mild winters.

Breeding biology

The breeding habitat is barren tundra. The nest is often established between stones or small tussocks, where the breeding bird is well camouflaged.

Breeds from the age of two or three.

The pair is monogamous and lays a single clutch of usually four eggs, which are incubated through 22 days by both sexes. The female departs from the territory after the eggs have hatched, leaving the chicks to the male, which leads them until they are able to fly after a good 20 days.

The oldest bird registered was 26 years and 8 months of age.

▲ The autumn migration of Red Knots culminates in August–September. Depicted here is a small group of Siberian *C. c. canutus* consisting of adults in breeding plumage and one juvenile at the front and one at the rear end. Öland, Sweden. 8.8. JL.

▼ Adult on nest.
The showy red breeding plumage may seem out of place when the Red Knot is roosting on pale sandy flats during the spring migration, but when brooding between lichens and rocks the spangled upperside merges well with the scant vegetation of the tundra. June. Taymyr, Siberia. JKAM.

Migration

Spring migration to the Arctic starts from the mid May to early June from the regular migration staging posts in the Wadden Sea, on western Iceland and Troms and Finnmark in north-western Norway.

The autumn migration begins with adult birds in early July and peaks towards the end of July to early August. Juveniles arrive from mid August and peak in mid September. Birds from Siberia as well as from the north-eastern Canadian population gather along the coasts of western Europe, primarily at the Wadden Sea. In September the Siberian birds leave and continue migration towards winter quarters in Mauritania and along the west and south-west African coasts.

The Canadian and Greenland birds winter primarily on the English east coast. They head back to the Wadden Sea in March–April to restore their fat reserves before the migrating to breeding places in May.

Distribution

Does not breed in Europe. This high Arctic species is distributed in northern Canada, in north and north-eastern Greenland, on the Taymyr in Siberia, the Wrangel Island and locally in Alaska.

▲ First winter, still with unmoulted, juvenile coverts and tertials.

On the migration staging posts and the winter quarters the Red Knot eats polychaetes and crustaceans, but also eats smaller bivalves, which are often swallowed whole. When low tide exposes the mussel flats, Red Knots can be seen feeding in their thousands, to rebuild the fat reserves before and after the long journey to Siberia and north-eastern Canada. 25.1. JL.

▼ The Red Knot is a hardy species, and smaller numbers of the Greenland and north-east Canadian population, *C. c. islandica*, winter each year at the Wadden Sea. On occasion they are also seen in small groups along other coasts. The two first winter birds seen here are associating with a group of the dark Icelandic subspecies of Common Redshank, *T. t. robusta*, which also winters in north-western Europe. The nominate subspecies of the Common Redshank winters in south-western Europe and west Africa. The Wadden Sea. 27.1. JL.

SANDERLING
CALIDRIS ALBA

Meaning of the name

'The white sandpiper'
Calidris, from Greek *kalidris* or *skalidris*, name for a greyish bird along the water, mentioned by Aristotle.

Alba, white.

Jizz

L. 20–21 cm. Ws. 35–39 cm.

The same size as Dunlin, but with a more rounded, short-necked appearance. Has a shorter, rather thick and straight bill and shorter legs. The only *Calidris* sandpiper with no hind toe.

Frequents deserted sandy coastal beaches, where it forages on small animals in the flotsam using a crouched stance with rapid, darting spells of fast runs forth and back between the waves in the surf.

In flight. Has a broad, prominent, white wing-bar and a conspicuous greyish-black central stripe on the white rump and greyish tail. Very social and often seen in flocks with Dunlins during migration.

▲ Adult breeding.
The rusty upperside, the short, almost straight bill, the broad wing-bar and the white rump with dark centre distinguish the Sanderling from all the other commonly occurring *Calidris* species in Europe. 25.5. JL.

Similar species

In breeding plumage it can be confused with Little Stint and Red-necked Stint, which both have a reddish plumage. Refer to these.

▼ Adult breeding.
Note the blackish-brown carpal joint, which is a species-specific key feature in the winter plumage. The carpal joint is often hidden by the breast feathers, however, as can be seen in the photos of non-breeding birds on the following spread.

The Sanderling is the sole *Calidris* species that has only three toes. The hind toe is absent, which is usually seen clearly on running birds.

The plumage is variable, from dark brick-red to almost lacking colour, as on this very pale individual. 26.5. LG.

Plumage and identification

Adult breeding. Male and female can only be safely told from each other on the breeding grounds. At a distance, an almost black-and-white wader with head, breast and upperside rufous and black with white fringes set against the pure white underside. The legs and the short, straight and robust bill are black. The intensity of the rufous in the plumage varies from darker rusty-red to reddish-yellow, as does the size of the black feather centres, which may make the bird appear more or less dark. Some individuals appear all pale, particularly at the onset of the season when the fresh feathers have broad pale fringes.

In flight. In all plumages the broad, white wing-bar and the black, central stripe of the tail are conspicous.

Adult non-breeding. Looks almost white with pale grey upperside and almost white head and white underside. The characteristic blackish-grey area on and around the carpal joint is a good field mark on birds on the ground, as long as the carpal joint is not hidden by the white breast feathers.

Juvenile. In fresh plumage the juvenile has a characteristic almost chequered black-and-white upperside as well as buffish breast sides.

Subspecies

Refer to Distribution.

Voice

The call is a single, hoarse '*chip*' or '*pit*'. From roosting flocks the many calls merge into a lively babble.

The song is a monotonous, croaking reel, '*dreet-dreet-dreet*', carried out in a rising and falling song flight with series of flapping wing strokes and glides on curved wings.

▲ Adult breeding.
In very fresh breeding plumage the new feathers have broad, white edges, which are gradually worn off during the spring and summer, thus giving the bird a darker appearance.

The rare Asian vagrant, Red-necked Stint, in breeding plumage closely resembles Sanderling, but it is smaller with more pure orange-red colour on the breast and neck, and it has hind toes. 26.5. LG.

▼ Adult breeding, in a more worn plumage with partly abraded white fringes and a more russet-red hue to the plumage.

Also note how fat the bird appears on the vent where a fuel supply of pure fat is being stored.

This surplus fat reserve helps the bird to survive its direct migration to the Arctic breeding places.31.5. DP.

Habitat

It is found primarily along the coasts of western Europe, preferring undisturbed sandy beaches with direct access to the sea and the surf zone where it feeds in small groups.

Breeding biology

Breeds at the age of one to two years on high Arctic tundra with scarce vegetation of dwarf willow, grasses and saxifrages, often near water with rich insect life.

The pair is monogamous, though exceptionally the female pairs with several males. Uses the same territory for years. One to two broods, possibly a third with a second male, are incubated for about 28 days by both sexes. In the cases of two broods, the parents each incubate one of them. The young fledge after 17 days and thereafter the parents leave the breeding area.

The oldest bird registered was at 18 years and seven months of age.

▲ Juvenile.
Told from adult in winter plumage by the black-and-white streaked mantle and the almost chequered pattern on the scapulars as well as the collar like rosy-buff wash on the breast sides. 4.9. JL.

▼ Post-juvenile.
Juvenile moulting into its first winter plumage with grey mantle and back feathers, but still with some retained juvenile coverts and tertials. 23.10. LG.

▼ The four eggs of the Sanderling in a hollow in a blanket of Greenland vegetation.
June. Danmarkshavn, north-eastern Greenland. LG.

Migration

Seen quite commonly during spring migration, from March to early June with peak numbers in middle to late May. At this time the birds, prior to departure to the Arctic breeding areas, can be seen in their beautiful rufous-black and white spangled breeding plumage. The autumn migration is initiated by the adults from the middle of July and culminates in the middle of August, when the juveniles also appear. Their numbers peak during the middle of September and migration draws to a close by early November.

Good numbers of birds winter along the shores of west and south-western Europe and by the Mediterranean, while the majority winter along the west African coast.

Distribution

The only population within the boundaries of Europe is found on Svalbard, which holds 50–100 breeding pairs. It is thus practically only seen on migration and as a winter visitor in Europe.

Sanderling is divided into two subspecies. The nominate form, *C. a. alba*, is distributed in Arctic Canada, Greenland and Svalbard to Severnaya Zemlya and Taymyr in Siberia, with winter quarters from western and southern Europe and Africa towards the east to southern Asia and Australia.

C. a. rubida breeds in north-eastern Siberia, Alaska and in northern Canada. Winters on east Asian coasts as well as from North America to southern South America.

▶ Adult non-breeding.
Note the dark carpal joint which peeks out above the breast feathers and is a diagnostic feature in the winter season.
This bird is of the North American/Asian subspecies *C. a. rubida*. 7.1. California. DP.

▲ Adult non-breeding. Has greyish, finely streaked crown, grey back and grey arm coverts as well as fresh flight feathers. In flight the species is distinguished from other *Calidris* species by the very dark forewing. 21.3. BLC.

PURPLE SANDPIPER
CALIDRIS MARITIMA

Meaning of the name

'The sandpiper of the sea'

Calidris, from Greek *kalidris* or *skalidris*, name for a greyish bird along the water, mentioned by Aristotle.

Maritima from mare, the sea.

Jizz

L. 20–22 cm. Ws. 42–46 cm.

Slightly larger than Dunlin and in all plumages generally grey with a plump body and a deep chest, which give it a tendency to appear front heavy and permanently out of balance.

Differs from all other smaller *Calidris* species by the slightly curved bill, which is deep yellow on the inner third and by the yellow, rather short legs.

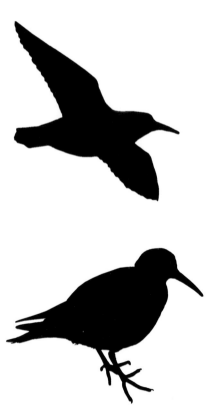

Roosting birds are very confiding, and because of camouflage you may not notice them until you are only a few metres away.

In flight. Appears short and chubby with a prominent body. At a distance it gives an almost black impression with a narrow white wing-bar on both arm and hand.

Plumage and identification

Throughout the year it exhibits the same greyish ground colour on the upperside, and a greyish-white underside with greyish to blackish-grey spots and streaks.

Adult breeding. The crown and face attain a rust-coloured tinge. Mantle and scapular feathers have rufous-buff to rusty-red edges. Adult non-breeding. Has uniform grey head and densely blotched neck and upper breast. The upperside is very dark greyish-brown with a purple sheen on the mantle and scapulars when seen in the right light.

Juvenile. The pure juvenile plumage is only seen in the breeding area, as the species arrives late to the north-western winter quarters. Refer to the captions and photos for other plumages.

Voice

The call is a short, soft 'tritt' or 'huit-wit'.

The song, which is performed in a display-flight with raised head and prominent breast is rapidly repeated trill 'pooh-rrhooh-ee-roohooh-roohoohee'. The display-flight begins with a circling ascent on stiff wings, and is ended by a fast descent in zigzag. The bird lands with wings raised in an 'angel like' 'V' as is also seen in, for example, Common Redshank.

Habitat

In the winter season seen in small numbers on rocky shores, piers, breakwaters and stony headlands. The birds occur singly or a few together, exceptionally in larger flocks, on apparently regular roosting and feeding places, from where visits to nearby coasts are frequent.

Breeding biology

Breeds on open, dry, stony tundra, but also on damp ground at low and high altitudes as well as on stony grounds near the coast. The pair is monogamous and solitary, preferring to use the same breeding territory for years. The single clutch of four eggs is laid in a lined depression in a mossy tussock and is incubated by both sexes for 22 days. Around the time of hatching the female leaves the brood and the responsibility to the male.

The young are able to fly after *c.* 24 days.

◀ Adult breeding.
The rufous-red fringes on the head, back and shoulder feathers are about the only variation in the overall blackish-grey plumage throughout the year.
25.5. Iceland. JL.

▲ Adult breeding. The species is extremely well camouflaged in mountainous habitat and on the tundra where the it often remains unseen until it moves a few metres from one's feet. The blackish-grey look and the mustard yellow colour on bill and legs separates it from all other European waders. 1.6. Iceland. DP.

▼ Juvenile. Birds in a fresh juvenile plumage like this one are seldom seen outside the area of the breeding distribution. The species is hardy and does not migrate towards the winter quarters in northern Europe until the autumn, when the moult into first winter usually has commenced. 10.8. Svalbard. SD.

▲ Post-juvenile.
Juvenile moulting into first winter, determined by the crown, which is still spangled in golden and, in particular, by the broad, white edges to coverts and tertials. 7.10. KBJ.

▼ Adult non-breeding.
Contrary to first winter, the adult has more diffuse edges to the feathers of mantle and scapulars as well as greyish, less clear cut edges to the coverts, giving an overall more uniform and grey look. 27.1. LG.

Migration

The Purple Sandpiper is a hardy bird, wintering further north than any other wader, even to southern Greenland. The migration routes and the specific winter quarters of the different populations were unknown until recently, and the species has been surrounded by mystery.

Within recent years, ringing studies with colour-rings and flagging, as well as placement of geo-locaters/light-loggers on the legs of the birds, has revealed some of the secrets.

The departure from the breeding grounds to nearby coasts is initiated by the females. There the adults moult the flight feathers, rendering them flightless for a period. In Iceland this happens in late June and later the females are followed by the males and the juveniles in the course of August. Depending on the origin of the population, the most important European winter quarters stretch from the British Isles to the west coast of Norway and as far south as to the French and Spanish coasts of the Atlantic. The return migration takes place from August to as late as November.

Recorded data show that the migration takes place via two routes at least; a western one along the Norwegian and Swedish west coast and an eastern one over land and down to the Gulf of Bothnia, possibly including birds from the Russian population.

The spring migration towards the breeding areas takes place from the middle of March to the middle of June.

▲ Non-breeding. In the winter season the Purple Sandpiper is seen singly or in small groups along stony coasts, piers and breakwaters, where it hops from stone to stone looking for amphipods and other small creatures, often alongside Ruddy Turnstones. In flight it appears a stocky, deep-chested and plump wader with broad wings. Distinguishing characters from Dunlin, which is also seen in the winter, are the very dark upperside with narrow wing-bars, and the dark underside with dark flanks and streaking on breast and belly. 23.1. KBJ.

Distribution

Breeds in Arctic and sub-Arctic climates from north-eastern Canada to the east over Greenland to Svalbard, and further eastwards along the coasts of the Arctic Ocean to Taymyr in Siberia. Additionally in Iceland and on the Faroe Islands as well as in northern Norway and Sweden.

Canadian birds and possibly some from west Greenland winter along the coast of eastern North America.

▶ First winter with worn, but still white edges on the coverts and a single tertial.
Note the faint purplish sheen, which is only seen in the winter plumage and has given the species its English name. 7.1. LG.

LITTLE STINT
CALIDRIS MINUTA

Meaning of the name

'The small sandpiper'

Calidris, from Greek *kalidris* or *skalidris*, name for a greyish bird along the water, mentioned by Aristotle.

Minuta, small.

Jizz

L. 12–14 cm. Ws. 28–31 cm

One of the seven stints and the smallest European wader. It is used as the reference for the other stints. The Little Stint is conspicuous by its tiny size. It has a small body and a short, black, almost straight bill and dark legs. Often seen alone or a few together, usually in association with Dunlins, alongside which it looks almost like a chick. Standing birds are very rounded in shape with quite long tails and long primary projections.

Forages actively with constant nodding, sewing machine-like movements. Juveniles have a white 'V' on the back, formed by the edges of the feathers of the back. This character is less prominent in the adults.

In flight. Can be told from the similarly sized Temminck's Stint by the uniformly grey tail feathers. On Temminck's Stint the outer tail feathers are white.

Plumage and identification

Adult breeding. Has rusty-orange upperside with black feather centres on the scapulars, which are tipped white when very fresh.

Standing birds show a more or less well defined whitish or golden 'V' on the back. This pattern is formed by the pale edges of the upper scapulars. The crown, nape and particularly the breast sides have black streaks. The short, almost straight bill is black. The cheek is rusty-orange and the throat as well as the eye-ring is white. The rest of the underside is white. Black legs.

In flight. Has a grey tail set against the white rump, with a narrow black centre. Narrow white wing-bar. Adult non-breeding. Has dull, greyish-brown upperside with streaked crown, nape and back as well as narrow, black shaft streaks on scapulars and coverts. The sides of the neck are dirty greyish, the rest of the underside is white.

Juvenile. Pale head with streaked crown, dark lores and a broad, pale supercilium. Patterned like an adult in breeding plumage, but lacking the rusty-orange hue. Instead the feathers of the upperparts have larger, black centres edged white and rusty-buff. The white edges of the mantle feathers form a distinct, white 'V'.

Similar species

Can be confused with Sanderling, Red-necked Stint and Temminck's Stint in breeding plumage. Sanderling, however, is considerably larger and more stocky, has a different behaviour and occurs

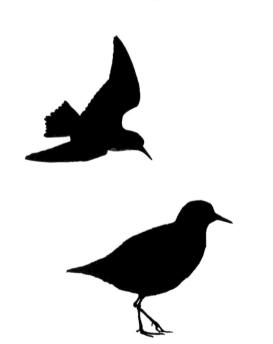

Adult breeding in fresh plumage. Differs in breeding plumage from the rare Asian vagrant, Red-necked Stint by, among other things, the white throat, the tertials with broad, rusty-red edges and the short primary projection, which barely extends beyond the tip of the tail. 20.5. Greece. DP

▲ Juvenile in flight with fresh, unworn feathers.
Easily told from other *Calidris* species and smaller waders by its tiny size, and from the similarly sized Temminck's Stint by the grey tail without white outer tail feathers. 11.9. Israel. LK.

primarily in smaller groups on quiet sandy beaches.

The very similar Red-necked Stint occurs as a rare Asian vagrant, and has a completely rusty-red neck in the breeding plumage.

The similar-sized Temminck's Stint has yellowish-green legs, a characteristic bill shape and an almost uniform, dull greyish-brown plumage throughout the year.

Refer to that species for further features.

Voice

The typical flight call is a hard, short, metallic '*pit*' or '*pit-pit*'.

The song is a monotonous and rapid '*tvi-tvi-tvi*', sometimes with a trilling '*tvirrr*'.

Sings from the ground or in butterfly-like song flight.

▼ Adult breeding Dunlins and an adult breeding Little Stint, which clearly stands out due to its size. 21.7. HS.

Habitat

The species is seen on spring and autumn migration at muddy and shallow coasts and inlets where it keeps to the shore. May also roost in suitable habitats at larger inland lakes. The Little Stint often occurs singly or in small groups, frequently in company with Dunlins.

Breeding biology

Breeds on coastal Arctic tundra.

The pair is monogamous or in complex relationships involving several males and several females.

The breeding biology is similar to that of Temminck's Stint: refer to this.

The clutch of four eggs hatches after about 21 days and the chicks are mostly cared for by one parent bird until they fledge. The oldest bird recorded lived for 14 years and eight months.

Migration

Both spring and autumn migration pass in a broad front overland in north-eastern and south-western directions respectively, to and from the Russian and Siberian breeding areas.

The spring migration takes place from mid May to early June, while the autumn migration is initiated by the adult birds from mid July to early August. The juvenile birds arrive in late August. Migrates primarily along the west coast of Norway, then further south to the secondary winter quarters in southern Europe and the primary winter quarters in Africa south of the Sahara. The large number of juveniles seen in some years is thought to relate

▼ Juvenile differs from most other small waders by its size and the characteristic white 'V' on the back. Furthermore, differs from Temminck's Stint in all plumages by the dark, blackish-grey legs.
3.9. Denmark. LG.

to fluctuation in the lemming population. After years with low numbers of lemmings the population of Arctic Foxes is correspondingly low, which allows better breeding success for waders.

The more eastern populations winter from the Arabian Peninsula to India and Myanmar.

Distribution

About 99% of the European breeding population of between 48,200 to 77,000 pairs breeds in Russia. Only 200–1,000 pairs breed in Norway and 0–5 pairs in Finland. Breeds in the Arctic part of northern Norway eastwards through Russia to eastern Siberia.

▼ Adult breeding to non-breeding.
The bird is in worn breeding plumage with very worn coverts, faded tertials and remiges and some retained rufous scapulars. The large, fresh feathers are the first newly emerged parts of the greyish-brown winter plumage. 28.8. Denmark. JL.

▲ Non-breeding.
Seen in winter in smaller numbers in southern Europe. The greyish-brown winter plumage is similar to that of Temminck's Stint, which likewise occurs in the area in winter.
 The two stints differ in leg colour; black in Little Stint and greenish-yellow in Temminck's Stint, and by the shape and colour of the bill, which is straight and black in Little Stint. Temminck's Stint has a slightly paler bill with a faintly yellowish-brown base and slightly upcurved edge of the lower mandible. 14.2. Oman. HJE.

TEMMINCK'S STINT
CALIDRIS TEMMINCKII

Meaning of scientific name
'Temminck's Sandpiper'
Calidris from Greek *kalidris* or *skalidris,* a greyish aquatic bird species, mentioned by Aristotle.

  Temminckii, named after the Dutch ornithologist Coenraad Jacob Temminck (1778–1858)

Jizz
L. 13-15 cm. Ws. 34-37 cm.
One of the seven small *Calidris* sandpipers known as stints (peeps in North America). The only sandpiper with white outer tail feathers and also the one with the most drab plumage. Size like Little Stint with a slightly upcurved lower mandible and yellowish-green legs. Stoops when feeding, often with slow, probing movements, and may resemble a miniature Common Sandpiper but does not bob its rear body up and down.

  A discreet and solitary bird.

▲ Temminck's Stint, non-breeding and Little Stint, adult non-breeding to breeding.
Both species appear drab greyish-brown in winter plumage, but notice the field characters: dark, yellowish-brown bill with a slightly upcurved lower mandible and greenish-yellow legs in Temminck's Stint; black, nearly straight bill and black legs in Little Stint. 11.3. Ethiopia. TVN.

▼ Non-breeding to breeding.
Stints can be difficult to identify, especially when backlit, but the upcurved lower mandible of Temminck's Stint distinguishes it from Little Stint. Also, notice the color of the bill, which when well illuminated is diffusely yellowish-brown, slightly paler towards the base, and the olive green to yellowish-green legs. 13.5. HS.

In flight. When flushed, it will often rise to considerable height and is distinguished from Little Stint, which is of similar size, by white outer feathers to an otherwise grey tail.

Similar species

Little Stint, which has a black, almost straight bill and very dark, sooty legs as well as uniformly grey tail feathers. See this.

Plumage and identification

Adult breeding. Greyish-brown upperparts with varying numbers of black centres to mantle and scapulars with yellowish or orange-tan fringes. The short bill is dark yellowish-brown; the lower mandible is slightly upcurved and paler yellowish at the base.

The crown has greyish-black and rusty streaks and the head has fine greyish-tan streaks, clearly contrasting with the white belly and greenish-yellow legs.

In flight. See caption.

Adult non-breeding. Uniform greyish-brown upperparts with narrow dark quills and greyish sides to the breast, contrasting with a white underside.

Juvenile. Easily distinguished from other young sandpipers by its characteristic upperparts with a dark sub-terminal band and golden-tan fringe on each feather, giving the birds a coarsely scaly appearance that is unique among small sandpipers.

Voice

The voice, which is a shorter version of the song, is a short '*prrree-tt*', not unlike that of a cricket.

The song, a loud metallic trill, is performed either from in display flight or perched on the ground.

▲ Adult breeding.
Male in display flight over its territory, here showing the four to six white or greyish-white outer tail feathers. Other sandpipers have uniformly grey tail feathers. 11.6. Norway. HS.

▼ Adult breeding.
Temminck's Stint has the drabbest plumage of all sandpipers, appearing greyish-brown in all seasons. However, the breeding plumage adds a bit of variation with streaked crown and mantle, and rufous scapulars with black centres. 22.5. Sweden. DP.

Habitat

During migration, may be encountered along muddy coasts, saline marshes, river deltas and a wide selection of inland biotopes, such as around lakes, along canals, and by ditches and reservoirs, where individuals or small groups may be found feeding.

Breeding biology

Breeds in mountain bogs and tundra as well as along rivers and bays along the coast and inland. Several pairs breed close together and each pair may be monogamous. However, they are often promiscuous, the female mating first with her own male and laying the resulting eggs and then mating again with another male and incubating the resulting eggs herself.

On rare occasions the second brood is left to male number two and the female proceeds to mate a third time and hatch this brood herself. Meanwhile, her first mate, before hatching his brood, mates with another female, whose mate is probably hatching her first brood.

The advantage of this behaviour is that the pairs utilise available resources to the maximum and, by hatching at least two broods in different nests, thus spreading the risk over a larger territory. For this to succeed the female must be in prime condition in order to produce at least eight eggs, totalling more than double her body weight. Also, hatching must coincide with the annual explosion of insect abundance, which is vital for the chicks' chances of survival.

Migration

Migration to and from Scandinavian mountains and Russian tundra covers a wide front across Europe. The species is scarce inland, though found more frequently than other sandpipers at gravel pits, lakes, marshes and other wetlands. Spring migration in Europe is from early April to late May. Autumn migration is from mid July to late September. Birds from Fennoscandia and probably northwestern Russia are scarce winterers along the coastlines of the Mediterranean, but primarily winter in Africa, south of the Sahara. Eastern populations winter from the Arabian Peninsula across India to south-east Asia.

Distribution

The European breeding population numbers 58,100 to 167,000 pairs. Only 12% of the population breeds in Sweden, Norway and Finland. The remaing 88% breed in European Russia. This species is distributed from northern Fennoscandia across northern Russia and Siberia to the Bering Strait.

▲ Juvenile in flight.
Note the coarsely scaly upperparts with golden fringes, and the white outer tail feathers. 22.8. NLJ.

▼ Non-breeding.
The drab, greyish-brown winter plumage of Temminck's Stint is similar to that of Little Stint, but Temminck's may easily be distinguished by its pale, greenish-yellow legs, its slightly upcurved lower mandible, and its four to six white outermost tail feathers. 30.3. Dubai. HS.

◀ Juvenile. The coarsely scaly upperparts with a characteristic dark sub-terminal band bordered by golden-tan fringes is unique among the seven species of stints, of which only Temminck's Stint and Little Stint are recorded regularly in Europe. Sweden. 2.8. DP.

RUFF
CALIDRIS PUGNAX

Meaning of scientific name
'The belligerent sandpiper'
Calidris from Greek *kalidris* or *skalidris*, a greyish aquatic bird species, mentioned by Aristotle.
Pugnax, belligerent, *pugnare*, to fight.

Adult male. Non-breeding to breeding.
In January to March the male Ruff commences partial moult from winter to breeding plumage. This includes head and body feathers, tertials, tail feathers and some wing coverts. This bird shows fresh black feathers on its head and neck as well as two secondary coverts with reddish markings. The spectacular ornamental ruff and head tufts are grown from late March until May. 26.2. Eilat, Israel. KBJ.

Jizz
Male: L. 26–32 cm. Ws. 54–58 cm.
Female: L. 20–25 cm. Ws. 48–52 cm.
The male in breeding plumage, with its ornamental feather collar, head tufts and warty facial skin, has the most spectacular appearance of all waders and is unmistakable despite considerable variation in plumage. Outside of the breeding season, female, male and juvenile are all similar; large sandpipers with medium length, slightly decurved dark or bicoloured bills, black-spotted, greyish-tan upperparts with a scalloped appearance and rusty hue, and orange to greenish-yellow legs. The wing coverts and back feathers have clear pale fringes. Alert birds frequently assume a 'giraffe-like' stance, with body diagonal, neck stretched and head lifted. It has a rather small-headed and small-eyed look compared to most similar-sized waders.

Feeds seemingly arbitrarily and often runs clumsily in a stooped posture.

Its featureless appearance may cause confusion, but in fact is an important characteristic in its own right, standing out in mixed flocks of waders.

In flight. Flies fast and easily, carried by long, wide wings, the body slightly curved. The head and rear body droop and the toes protrude beyond the tail. Alighting is preceded by a characteristic long glide and often ends in a short, fast run. The upperparts are mottled greyish-brown and it has a characteristic white 'V' or horseshoe marking to the rump.

Plumage and identification
Adult breeding, male. Usually visibly larger than females, but is mainly distinguished by its unusual

▲ Non-breeding, female.
The combination of a hunch-backed appearance in flight with slightly drooping wings and a white rump marking is not seen in other species and distinguishes this species with certainty in mixed flocks of waders. Note the slimmer, shorter body compared to the male in the photo to the left. 29.3. Eilat, Israel. EFH.

▼ Adult breeding, female.
The female is the size of a Common Redshank but has a uniformly dark bill and orange legs.
   The colours of her plumage match those of the surrounding meadow and tundra, where she alone builds the nest, and hatches and rears the chicks until fledged. 5.5. LG.

ornamental feather collar – the ruff – and the prominent head tufts. There is considerable variation in colour of the head feathers, back feathers and scapulars, from black and metallic purple to nearly pure white. Between these is a multitude of colour variations, including brownish-red, orange and cream, decorated with bars or specks.

The bare facial skin between the eye and the bill is warty-looking, mostly yellowish or reddish, and the bill may be dark, orange-red or pink, often darker towards the tip. The legs are usually orange-grey, bright orange or reddish-brown. Some males resemble females in breeding plumage.

Female. Usually visibly smaller than the male. Upperparts are greyish-brown, appearing scaly with black feather centres bordered by pale rims on mantle, scapulars and back feathers, and a variable amount of rufous bars, especially on the very long tertials, which are characteristic for this species. The bill is uniformly black; the breast is variably streaked in greyish-brown to rufous extending to the flanks. The remainder of the underside is whitish and the legs bright orange.

Adult non-breeding. Large, pale-rimmed greyish-brown feathers and tertials with rufous barring dominate the upperside. The bill is usually bicoloured, greyish to pink at the basal half and dark towards the tip.

Neck, head, breast and flanks are mottled in greyish tan. The belly is greyish-white, the legs pale orange. Pale males have partly or completely white heads and necks.

Juvenile. The feathers of the upperparts appear characteristically coarsely scaly with wide cream fringes. The crown is black-streaked, the bill greyish-black, the neck and breast cream to pale tan contrasting with the greyish-white belly and it has greenish-yellow legs.

See captions for further plumage and flight characteristics.

▲ Adult breeding, males and females.
A mixed flock of Ruffs on spring migration. The upperside of the wing is generally grey with a narrow white band along the greater wing coverts. There is a characteristic V- or U-shaped marking of the rump, depending on whether the central upper tail coverts extend to the tail. Furthermore, the toes protrude well beyond the tip of the tail.
   The males are readily distinguished by size and coloration of back, head and neck. 24.4. NLJ.

▼ Typical outlines of Ruffs at a lek during spring migration.. The number of roosting birds peaks in mid May when small groups may often be seen dancing spontaneously, with some females initiating mating. 4.5. LG.

Voice

Is very quiet all year round, with no song.
The call, a hoarse, rapid '*kwe-kwe-kwe*', is heard as a warning at the breeding grounds and as a contact call between birds in groups.

A surprised, soft '*huw*' is rarely heard and mostly when birds are flushed.

Habitat

Their preferred breeding habitat is wet meadows and marshes with short vegetation and, towards the north-east, tundra.

Ruffs are common on spring and autumn migration when individuals and small groups may be seen along the coast, on sandy or muddy ground and inland on wet meadows and near lakes and brooks. In autumn they often join groups of European Golden Plovers roosting on meadows and ploughed or stubble fields.

▲ Adult breeding.
A male in full ornamental plumage with its ruff and head tufts, which are fluffed out only at the lek. 5.5. JL.

▶ Adult breeding.
A male checking a female's willingness to mate, at a short stop-over at a roost during spring migration across the Wadden Sea. 5.5. LG.

▲ The breeding plumage of males varies from near white to black. The ornamental feather collar and head tufts are only extended at the lek, where the darkest males dominate and paler individuals occupy the periphery. 3.5. JL.
▼ Notice the long tertials and drooping scapulars, which are typical for this species. 5.5. LG.

Breeding biology

Ruffs may be observed dancing spontaneously at roosts during spring migration and may even mate.

This species is opportunistic, generally migrating towards traditional leks, but when conditions are unsuitable, for example because of changes to water level or vegetation, or late snow cover, the birds will readily find new, more suitable locations.

No song – just dance

Ruffs breed from the age of two but do not form pairs and have no song.

Instead, the males have a spectacular breeding plumage and display dance.

When they arrive at the breeding grounds, the males establish a lek, usually on an elevated spot with short vegetation, allowing them to fight and display their ornamental feathers for all to see.

Three dominant types of males

The darkest males occupy the centre of the lek where they each maintain a small territory, which they defend with threat postures and air duels with intense kicking and pecking.

The periphery of the arena is occupied by younger birds and dethroned males that are too weak to maintain a territory.

The lek is bordered by the palest, often white-ruffed males, so-called satellite males. After a complex exchange of mutual respect, threatening and evasive posturing, the males gradually accept each other's presence and avoid unnecessary fights, seemingly aware that all contribute to attract females.

There is no clear evidence that females prefer one specific male plumage variation to another. However, the white males do seem to act as beacons, visibly advertising the lek at great distance.

Flurry and stance

Whenever a female alights at the lek the males puff up their collars and head tufts in a flurry and flap their wings. The dance swiftly alternates between circling step dance, upright military stance, sprints and lounging at rivals. The males closest to the female freeze, crouched with spread wings, body inclined forwards, the bill pointing at the ground, possibly to appease the female who is free to pick her choice among the different types of males. The dark, dominant males at the centre of the lek, however, are possibly the most popular.

Males in drag

A surprising trait in the Ruff is the existence of so-called faeder males (Friesian or Old English for 'father'), which constitute 1% of the population. These closely resemble overgrown females, lacking head ornamentation. They move around undetected at the lek and 'steal' mating opportunities by rapidly moving between the chosen male and the crouching, ready-to-mate female, mating with her in an instant. That the females sometimes accept this shows that the strategy is a successful alternative for male Ruffs.

Considering that their testicles are larger than their brains during the breeding season, it is no surprise that males are obsessed with mating during this period.

Single mothers

After mating, the female alone rears the brood. The nest is built in a tuft of grass or other tall vegetation and is lined by a bit of plant material. The single clutch, four eggs, hatches after *c.* 23 days. This is one of the few waders to actively feed its chicks, the female bringing insects for them during the first few days. After about 28 days the brood is fledged;

the female leaves immediately before or after this.

The males have played their part and dancing ends around the time when the first chicks are hatched. At this time, the males start moulting their ornamental feathers and commence migration southwards towards the winter quarters, thus avoiding the need to compete for food with females and chicks.

The oldest known ringed bird was 13 years and 11 months old.

Migration

Numerous on spring migration from early March to early June, en route to its breeding grounds in Fennoscandia and western Russia. Most of the Scandinavian population roosts at the Wadden Sea, whereas eastern populations migrate overland further eastwards.

Autumn migration covers a wide front across Europe and is initiated by adult males in late June, followed by females from mid July. Both sexes quickly

Though males dance and fight in a flurry of feathers, occasionally even drawing blood, it is still the female that picks her choice among them. Males will crouch motionless on the ground at the lek as soon as a female is within range. On the other hand, the female alone manages nest-building, incubation and parental care. 18.5. Finland. JP.

migrate to their primary winter quarters in Africa, south of the Sahara. During August migration of juvenile birds peaks and decreases during September. Individual juveniles are reported until November.

Small numbers of Ruffs winter in western Europe and around the Mediterranean.

Distribution

Judging by the number of displaying males – 265,000 to 1,650.000 – the European population is estimated at 531,000 to 3,310,000 adult individuals. Some 94% of the European population breeds in Russia, whereas Norway, Sweden and Finland together total around 5%.

The species is in decline in the south-western part of its distribution, which is moving eastwards due to cultivation, drainage and no or too intensive grazing of meadows and, possibly, change of climate.

This species breeds from north-western Europe across Scandinavia and the Baltics through parts of north-eastern Europe, Russia and Siberia to the Bering Strait. Eastern populations winter from Indian coastlines to southern Asia.

Immatures may be dark, pale or rufous, but always have large dark-centred scapulars and wing coverts. Common to all feathers is a wide pale cinnamon or cream fringe. The bill is greyish-black and the legs are greyish-green to greenish-yellow.

▲▶ Juvenile, presumably a male due to its triangular head shape and its feet protruding only little behind the tail.
The white marking on its rump is 'V' shaped, divided by a dark central bar. In other individuals it may be shaped like a horseshoe and resemble the white rump of Curlew Sandpiper. 5.8. JL

▶ Post juvenile, presumably a male with its heavy body, triangular head shape and long legs. Notice the lesser secondary coverts, which have already moulted and been replaced by the grey feathers of its first winter plumage. 24.9. LG

▶ Breeding to non-breeding, male. As soon as mating is concluded the males leave the leks and females and head for Africa. The showy feather collar and head tufts are moulted first when the greyish-brown winter plumage replaces the breeding plumage.
Oman 2.9. HJE.

▼▶ Adult non-breeding. Ruff resembles Common Redshank in winter plumage, but is easily distinguished by its long tertials, drooping feathers and the white area around the base of its bill.

Leg colour may be greyish, greenish orange-yellow or yellow.

It is a challenge to determine the sex of single birds. The male is usually 20% larger than the female, but size is variable in both sexes. Judging by the size of the Common Redshanks nearby, this bird is probably a male. 27.1. Holbæk Harbour. LG.

▼ Non-breeding. At their winter quarters, flocks of Ruffs consisting of purely males or females are often seen. Judging by size and head shape, this flock consists exclusively of males in winter plumage.

Age determined by new, dishevelled feathers, a dark bill, yellow towards the base and orange-yellow legs as seen in the bird to the left.

Notice the very small individual to the right. This may be a solitary female or one of the female look-alike faeder males. See Breeding biology. 11.11. Salalah, Oman. KBJ.

Juvenile Ruddy Turnstone with Red Knots and Dunlins.
  Ruddy Turnstones in all age groups and plumages can be identified by the distinctive, white markings on the upperparts, which form this species' unique pattern.
  Only the Black Turnstone has a similar pattern and this species has never been recorded in the Western Palearctic. 4.9. HS.

The turnstones belong to their own genus within the family Scolopacidae. There are just two species, both of which breed in the Northern Hemisphere.

The rare Black Turnstone is found in north-western North America. The Ruddy Turnstone has a circumpolar breeding distribution, including southern Fennoscandia.

The turnstones resemble sandpipers, but have shorter legs and are muscular with short necks, a wide breast and a characteristic, short, wedge-shaped bill, which is used as a lever when they turn over stones and sheets of seaweed in search for amphipods and other small creatures.

Outside of the breeding season, the Ruddy Turnstone is a long-distance migrant and a cosmopolitan species that may be encountered at most coastlines around the globe.

Due to its similarity with medium-sized sandpipers, Ruddy Turnstone has been placed here, after Ruff and before the plates with rare American and Asian sandpipers.

RUDDY TURNSTONE
ARENARIA INTERPRES

Meaning of the name

'The sand-dwelling treasure hunter'
Arenaria, related to sand; *arena*, sand.

Interpres from *inter*, between and *pretium*, value or reward. Named after this species' habit of poking its bill between stones and turning them over, in expectation of reward.

Jizz

L. 21–26 cm. Ws. 50–57 cm.

A compact, powerfully built wader with a short neck, a short, powerful, wedge-shaped, greyish-black bill and short orange legs.

Unmistakable in its black, white and orange breeding plumage. Feeds in a unique manner with lowered head and bent legs, flipping even quite

Adult breeding. Presumably a male, sporting bold, intense colours with no brownish smudges to the white, and with distinctly black-streaked crown. The white wedge-shaped back may be seen in several waders, especially *Tringa* sandpipers, but this particular combination of white patches on the upper wing, back and tail is unique to Ruddy Turnstone and makes it an easy bird to identify at any time of the year. 10.5. JL.

large rocks and sheets of seaweed to snatch up any exposed sandhoppers and other small creatures.

In flight. The body is compact and short on long, wide wings with characteristic black-and-white markings on the upperparts. It has white wing-bars, a narrow stripe of white on the back and a diagonal white wedge at the base of each wing.

Plumage and identification

Adult breeding, male. The crown is streaked black and white, the nape white. Mantle, scapulars, median wing coverts and tertials are intensely rufous. The breast is black, contrasting with the white underside.

Legs are orange-red all year round.

In flight. The characteristic black-and-white pattern of the upperparts distinguishes this species from other waders the same size, particularly sandpipers. See captions for details.

Female. Brownish-black crown and nape markings and less intense rufous coloration. Overall, has a subdued, smudgier appearance due to a brownish wash on the white parts of its plumage. The sexes are best distinguished at the breeding grounds.

Adult non-breeding. Brown head and all black and rufous markings on the body replaced by greyish-brown.

Juvenile. Like a faded version of the breeding plumage, without rufous, but with heavily streaked crown and typical juvenile feathers bordered by wide, pale fringes to scapulars and cream

fringes to wing coverts and tertials. In flight displays the same characteristic upperside markings as adults.

Voice

The call is a hard '*kie-kie-kiew*' and a softer '*hi-hi-hi*'.

The song is performed from song flight with varying sequences of slow, rowing wingbeats interspersed with gliding on curved wings. It is an explosive, rapid and hard, metallic trill, '*bik-biik-bika-gegegegegegege*'.

Distribution

In Europe the Ruddy Turnstone winters along the coastlines of western Europe and the eastern Mediterranean. On spring and autumn migration this species may be encountered singly, in pairs or in small groups, on shingle and rocky coasts, groynes and piers and further inland near lake shores and rivers. Outside of the breeding season the Ruddy Turnstone is one of the waders with the widest distribution, visiting most of the planet's ice-free coastlines.

▲ Note the black-and-white upperparts with white markings on wings, scapulars, back and tail, forming a set of flight characteristics unique to the Ruddy Turnstone. This mixed group of waders also includes Red Knots, Dunlins, two Common Ringed Plovers (blurred in the foreground) and to the right four Sanderlings in adult breeding plumage with their characteristic wide, long, white wing-bars. 24.5. JL.

▼ Adult breeding, male and female. The breeding plumage varies with some birds near monochrome and others very rufous. The sexes are easiest to determine when a pair is seen together. The female to the right is identified by slightly paler coloration and the smudgy head and nape, while the male shows more intense colours, pure white areas and a black streaked crown. 16.5. HS

▼ The Ruddy Turnstone makes its nest in a slight indentation in the stony tundra ground, where the camouflage of the eggs conceals them from gulls, skuas, ravens and Arctic Foxes. Danmarkshavn. Northeast Greenland. Juni. LG.

Breeding biology

This species breeds at two years of age, is monogamous and fiercely territorial. In Scandinavia the Ruddy Turnstone usually breeds alongside terns and small gull species, for better protection against egg robbers. The nest is often built under the shelter of a large stone, or within a domed tussock with a side entrance. In the tundra the nest is a simple hollow, sometimes lined with plant material. It is mainly the female that incubates the single clutch of four eggs, for about 24 days. After they are hatched both parents are involved in rearing the chicks for the first couple of weeks, after which the female leaves the breeding area. When the chicks fledge after about 24 days, the male soon leaves them too.

The oldest registered ringed bird was 21 years and five months.

Migration

Spring migration towards the breeding grounds is underway from early April to early July, along coastlines and over land. The breeding period is early May to early August.

Autumn migration through western Europe is begun by adult birds in early July, peaking in mid August. These are birds from northern Europe, and presumably small numbers from populations in north-western Russia. Juveniles do not arrive until August. Birds migrate via the Bothnian Bay and the Baltic Sea, southwards along the coastline of western Europe towards west Africa. At the Wadden Sea many birds linger until the end of the year. Throughout autumn they are supplemented by Canadian birds and usually some from Greenland

Juvenile Ruddy Turnstones fighting for a feeding territory at Blåvandshuk at the Wadden Sea. At the breeding grounds adults fearlessly pursue would-be egg thieves, such as large gulls and crows. Note the drooping tertials on the right-hand bird, revealing flight feathers that they normally cover and protect. 25.8. BLC.

too, which normally winter in the British Isles.

The primary winter quarters for European birds is the south-west European Atlantic coast and the African coastline up to Cape Town, South Africa.

Birds from eastern populations in northern Russia and Siberia are presumed to migrate across the Black and Caspian Seas towards their winter quarters in the eastern Mediterranean, the Red Sea, the Persian Gulf, and along the coastlines of the Indian Ocean all the way to South Africa.

Distribution

The European breeding population numbers between 15,900 and 37,100 pairs. The largest populations are found in Norway, Russia, Sweden and Finland. Denmark is the southern limit for the Arctic breeder and this limit is expected to move northwards as the climate warms.

Apart from the northern European population, the Ruddy Turnstone has a circumpolar distribution around the Arctic where it breeds on stony coastal plains, marshy valleys and tundra. There are two more subspecies in addition to the nominate form; one in Alaska and one in Canada.

▲ First winter, identified by juvenile wing coverts with pale fringes and white-edged tertials. Note the fresh, greyish-black scapulars.

Adult non-breeding has uniform greyish-black upperparts with pale edges to the wing coverts and is difficult to distinguish from first winter. 7.11. HS.

▼ Juvenile in the process of turning a stone, using its muscles in its short neck and its flat, wedge-shaped bill, which stabs, turns and twists when birds hunt for sandhoppers and other small creatures. This species will also feed on carrion. 27.9. LG.

Adult breeding.
A couple of North American Stilt Sandpipers at the breeding grounds in Canada. 8.6. DP.

227

American and Asian vagrants

Each year Europe is visited by variable numbers of rare *Calidris* sandpipers from North America and Asia.

These rare vagrants may turn up in spring but are most common in autumn. At this time, the number of birds migrating is greatest due to the addition of juveniles. These young birds are also more likely (because of inexperience) to become lost while migrating.

There are several possible reason for these rare visits. Birds may have been blown off course, have set off in the 'wrong' direction because of genetic abnormalities (reverse migration), or they may have simply have tagged onto a group of Dunlins, following these birds' normal route towards western Europe.

In order to avoid unnecessary page flipping in search for common species, the rare vagrants have been grouped here at the end of the *Calidris* sandpiper section.

Among the most tricky vagrants to identify are the group of seven closely related small sandpipers known as stints (or peeps in North America), of which only two regularly visit Europe, namely:

Little Stint, *C. minuta*.
Temminck's Stint, *C. temminckii*.

The remaining five species recorded in Europe are:
Western Sandpiper, *C. mauri*.
Semipalmated Sandpiper, *C. pusilla*.
Red-necked Stint, *C. ruficollis*.
Least Sandpiper, *C. minutilla*.
Long-toed Stint, *C. subminuta*.

WESTERN SANDPIPER
CALIDRIS MAURA

Meaning of the name
'Mauri's sandpiper'
Calidris from Greek *kalidris* or *skalidris*, a greyish aquatic bird species, mentioned by Aristotle.

Named after the Italian botanist Ernesto Mauri (1791–1836).

Jizz
L. 14–17 cm. Ws. 28–37 cm.
One of the seven stints.

In breeding, plumage resembles a miniature Dunlin without a black belly patch, but in its place dark brown arrowhead markings.

This species has the longest bill of the stints and is one of only two that show partial webbing between its forward-facing toes (the other is Semipalmated). Perched birds show little or no primary projection beyond the tail, depending on wear.

In flight. Narrow, white wing-bars, white rump with dark longitudinal central bar and toes that reach the tip of the tail or slightly beyond.

Similar species
In non-breeding and juvenile plumage, it resembles its American relative, Semipalmated Sandpiper. The female of the latter may have a bill of the same length and shape, making it impossible to distinguish these two species in the field.

Plumage and identification
Adult breeding. Head pale and brown-streaked with a slightly decurved and attenuated black bill. Has rufous ear-coverts, and heavy brown streaking on the crown.

The upperparts are speckled in blackish-brown and rufous, the black-centred scapulars with pale fringes. The heavy, blackish-brown streaks on the neck merge into dense brown arrowhead markings, which diminish to scattered specks on the flanks and the remainder of the white underside.

Legs are black to blackish-green with partial webbing between the forward-facing toes.

The female has a slightly longer bill than the male.

Voice
The flight call is a vibrating, brief '*jeet*'.

Habitat and migration
A very rare vagrant from North America or Siberia. Recorded from France, Ireland, Portugal, Spain, Sweden and the United Kingdom.

Outside of the breeding season may be seen on sandy or muddy coastal habitats and sometimes fens, lakes and river deltas close to the coast. Winters along the coastlines of southern North America, Central America and northern South America.

Distribution
Breeds on tundra in the western Alaska and eastern Siberia.

▼ Adult breeding to non-breeding.
Resembles a miniature Dunlin without a black belly patch, which is replaced by scattered brown arrowhead markings on belly and flanks.
Through the worn breeding plumage, the first grey winter feathers can be seen emerging on mantle and scapulars. California. 8.8. DP.

▶ Post juvenile.
Resembles a combination of Dunlin and juvenile Little Stint.

The bill shape resembles Dunlin's, but body, leg colour and the vague white 'V' on its back as well as the rufous scapulars recall juvenile Little Stint. This bird is moulting to its first winter plumage with fresh grey feathers on the mantle and a couple between the uppermost scapulars.

Useful field characters when distinguishing it from the closely related Semipalmated Sandpiper, *C. pusilla*, are bill shape, rusty scapulars, the central ones of which have black anchor markings (1), and the short primary projection. 8.8. California. DP.

▼▶ Non-breeding.
Western Sandpiper and Semipalmated Sandpiper, two rare vagrants from Alaska/Siberia and North America respectively, are near identical in winter plumage, and variation of bill length in both species may render identification extremely difficult. However, note that Western Sandpiper resembles a miniature Dunlin with slightly decurved, attenuated bill, whereas Semipalmated Sandpiper is more akin to a Little Stint with a more powerful, straight, blunt-tipped bill. 30.12. Florida. DP.

▼ On solid ground the webbing of the forward-facing toes may be seen. This distinguishes it from all other stints except Semipalmated Sandpiper, which has similar webbing. 30.12. Florida. DP.

SEMIPALMATED SANDPIPER
CALIDRIS PUSILLA

Meaning of the name
'The tiny sandpiper'
Calidris from Greek *kalidris* or *skalidris*, a greyish aquatic bird species, mentioned by Aristotle.
Pusilla, from *pusillus*, tiny.

Jizz
L. 13–15 cm. Ws. 34–37 cm.
One of the seven stints.

A very small sandpiper, resembling a Little Stint, but can be told from this and all other stints except Western Sandpiper by the partial webbing between its forward-facing toes. The bill, which is black and straight, is deep at the base with a blunt tip.

Note that longer-billed individuals' bills have a slightly curved tip, like that of Western Sandpiper.

On perched birds the tips of the primaries reach or do not quite reach the tip of the tail. In flight. Narrow white wing-bars and white rump with a dark at the centre.

The toes do not extend to the tip of the tail.

Similar species
Western Sandpiper, juvenile Little Stint and juvenile Red-necked Stint. See these and captions here.

Plumage and identification
Adult breeding. Overall grey and black-spotted appearance. Crown heavily streaked in black and yellowish-tan, a rusty wash to the ear-coverts and greyish-brown upperparts with black-centred scapulars containing a few reddish-yellow patches.

The upper breast and sides of the breast are more or less covered in coarse greyish-brown streaks, which may create a breast-band. The remainder of the underside is white with faint or no barring on the flanks.

Legs are black to blackish-green with partial webbing between the forward-facing toes.

Note that females on average have slightly longer bills than males, and that the bill may be slightly curved at the tip.

Adult non-breeding. Streaked crown, more or less distinct pale supercilium and uniform greyish-brown upperparts with diffusely darker feather centres. The sides of the breast are streaked, which may create a narrow breast-band.

Juvenile. Juveniles have a subdued appearance, drab and greyish-brown. Very similar to juvenile Red-necked Stint and boldly marked birds may resemble juvenile Little Stints. See captions for further details.

Adult breeding. The rather plain plumage is common to adults and juveniles. Some birds in breeding plumage are paler than this one, others look more like Western Sandpipers with more rufous coloration to crown and scapulars.
However, the breast and belly lack the bold, blackish-brown arrow markings of Western Sandpiper. Yukon, Alaska. 28.6. GV.

Voice

The contact call is a mellow, rolling '*uuituituituituit*' and the most frequently heard, characteristic flight call is a rasping, short '*chip*'.

Habitat and migration

A rare vagrant from North America. Has been reported from Albania, Denmark, France, Germany, Iceland, Ireland, the Netherlands, Portugal, Spain, Sweden and the United Kingdom.

Outside of the breeding season frequents sandy and muddy coastal localities and sometimes at similar habitats near fens, lakes and river deltas close to the coast. Winters in Central America and along the coastlines of northern South America.

Distribution

Breeds sparsely in eastern Siberia and commonly in the Arctic and low Arctic belt across the North American continent.

▲▶ Juvenile resembles a pale, washed-out version of juvenile Little Stint with a faint, pale 'V' on its back and a hint of of cream fringes to scapulars, wing coverts and tertials.

In fresh plumage, as seen here, it has a greyish-tan wash on the breast. The streaking may sometimes form a breast-band, and there are variable greyish-black, diffuse anchor-shaped markings on the lowermost wing coverts.

The legs are dark greyish-green.

See Western Sandpiper and juvenile Red-necked Stint for comparisons. 8.8. California. DP.

▶ Juvenile. An older, bleached, longer-billed bird than the one at the top, suggesting that it is a juvenile female, possibly from the eastern Canadian population, which on average have longer bills.

In Europe a single bird with nothing to compare it to may resemble a juvenile Sanderling. However, the latter is larger and heavier. 17.10. New Jersey. HS.

RED-NECKED STINT
CALIDRIS RUFICOLLIS

Meaning of the name
'The red-necked sandpiper'
Calidris from Greek *kalidris* or *skalidris*, a greyish aquatic bird species, mentioned by Aristotle.
    Rufus, red, *collis*, necked.

Jizz
L. 13–16 cm. Ws. 29–33 cm.
One of the seven stints.
A small, slender, long-tailed and long-winged stint with a short, black, straight and blunt-tipped bill, and black legs. Very similar to Little Stint, but without the pot-bellied appearance of that species.

Perched birds appear elongated and slender at the rear, with long tails and primary projections.

In flight. Narrow, white wing-bars and a dark longitudinal central bar to the white rump and grey tail. The toes do not project beyond the tail-tip.

Similar species
Little Stint, see this and captions.
Distinguished from other stints by the combination of black legs and the lack of webbing to the forward facing toes.

Plumage and identification
Adult breeding. The bill is black, short and straight with a blunt tip. The crown has dark streaks with a tawny wash at the rear. Cheeks and neck are tawny and contrast with the upper breast with its arrow-shaped, black specks, which create a near complete breast-band. The remainder of the underside is white and the legs are blackish-grey.

The upperparts display dark-centred scapulars with tawny markings above and grey below contrasting with their pale fringes.

Some tertials are tawny. Note that the secondary coverts are grey and contrast distinctly with the mottled upperparts. This is not seen in adult or juvenile Little Stints, where the secondary coverts are dark-centred with reddish-brown fringes.

Juvenile. Rusty markings on mantle and uppermost scapulars, typically contrasting with grey, pale-rimmed secondary coverts. The head is greyish-white with a streaked crown. The throat is white, contrasting with slightly streaked sides of the breast and a smudgy, diffuse breast-band.

Non-breeding. Birds in winter plumage are near identical to Little Stints and may be impossible to distinguish. However, Red-necked generally appears lighter grey with a white centre to its breast, not greyish-brown as in Little Stint, which often has a heavy or diffusely barred breast-band. For identification, general appearance, voice and habitat are important factors to consider.

Adult breeding, fresh plumage.
Red-necked Stint is the east Asian counterpart to the Little Stint, which it resembles in adult breeding and adult non-breeding plumages. Birds in full breeding plumage show a richer rufous colour than Little Stint and resemble miniature Sanderlings. In breeding plumage may be distinguished from Little Stint by uniformly grey secondary coverts devoid of dark centres and, as seen here, its orange-toned sides below the lowermost scapulars. Also has uniform grey tertials with long pale fringes, a few of these with a rufous wash. The tertials of Little Stint are blackish-brown with wide, brick-red fringes. Finally, this species generally lacks Little Stint's well-defined pale 'V' marking on its back. 3.5. China. DP.

Voice

The flight call is a short, whistled '*ti-rrll*', very different from Little Stint's hard, metallic '*pit*' or '*pit-pit*'.

Habitat and migration

A rare vagrant to Europe from Siberia. Recorded from Belgium, Denmark, Finland, France, Germany, Italy, Kazakhstan, the Netherlands, Norway, Sweden and the United Kingdom.

Migrates across land as well as along the coast to its winter quarters and is often encountered on mudflats, brackish lagoons and near large inland lakes, rarely on sandy and stony beaches.

Winters from eastern India southwards to Australia and New Zealand.

Distribution

Breeds in the low Arctic part of the tundra in the central and eastern part of northern Siberia, and sporadically in Alaska.

▲▶ Post-juvenile.
The first pale grey winter feathers can be seen on the mantle and under the uppermost row of the rusty-rimmed, black-centred scapulars, the only contrasting feature to an overall greyish appearance. This distinguishes this species from juvenile Little Stint, which has a far more colourful reddish-tan, black and white plumage with a distinct, white V-shaped marking on its back. See Little Stint for comparison. 3.10. China. DP.

▼ Adult breeding to non-breeding.
A very ragged bird with the remains of reddish scapulars from its worn breeding plumage. Albeit bleached, the neck is still pure brick red, which distinguishes it from Little Stint in similar plumage. Also, the uniform grey secondary coverts with greyish fringes are clearly visible.
Little Stints in breeding and juvenile plumages have dark-centred secondary coverts, with brick-red and reddish-tan edges respectively.
Note that there is no webbing between the forward-facing toes, a reliable feature to distinguish it from Western and Semipalmated Sandpipers in any plumage. 18.8. China. DP.

LEAST SANDPIPER
CALIDRIS MINUTILLA

Meaning of the name
'The very small sandpiper'
Calidris from Greek *kalidris* or *skalidris*, a greyish aquatic bird species, mentioned by Aristotle.
Minutilla, very small.

Jizz
L. 13–15 cm. Ws. 28 cm.
The smallest of the seven stints.
One among three small sandpipers with pale legs.
The smallest of all waders.
The bill is black, slightly decurved and pointed. The legs are orange-yellow to greenish-yellow, medium length and with long toes.
Perched, relaxed birds have a nearly spherical appearance. Often stoops and crouches with partly folded legs while feeding, making it appear even smaller.
In flight. Narrow, white wing-bars and a dark, longitudinal bar running along the centre of its white rump and grey tail. The toe tips reach the tip of the tail.

Similar species
Temminck's Stint and Long-toed Stint, the other two stints with pale legs. See these.

Plumage and identification
Adult breeding. A small and very dark sandpiper. Crown strongly striped in black and reddish-brown, face brownish with a greyish stripe behind the eye. Has a slightly curved and pointed black bill.
The black mantle with reddish-brown stripes is framed with a pale 'V' on the back. All scapulars have large, black centres, the uppermost with rusty fringes, the lower ones marked with rusty-beige and white. Shows fine white feather fringes in fresh plumage, which are rapidly worn away.

Note that the tertials have cinnamon-coloured, wavy fringes. This distinguishes this species from Long-toed Stint, in which the fringes are straight-edged.
The breast is heavily barred with narrow, black 'pen strokes' on the lower flanks. The remainder of the underside is white, and the legs are orange-yellow with long toes, though these are not so long as in the similar Long-toed Stint.

Voice
The flight call is a soft, rising, high-pitched *'proohoohit'*, similar to the short trill of a cricket.

Adult breeding.
The dark head, black bill and very dark upperparts with large, black feather centres generally makes it easy to distinguish from the other three small, pale-legged stints. Furthermore, it has wavy tawny markings on the tertials (1) and upper tail coverts. This is not seen in the closely related Long-toed Stint from Asia, which also has a greenish-yellow base to its lower mandible. (This bird has mud on its lower mandible.) 6.6. Canada. DP.

Habitat and migration

A rare vagrant from North America. Recorded from Belgium, Finland, France, Germany, Iceland, Ireland, Portugal, Spain and the United Kingdom. Frequents mudflats along the coast and inland lakes and ponds. Winters in southern North America, Central America and northern South America.

Distribution

Breeds across a wide belt across North America, through the low Arctic zone to the tree limit.

▶ Juvenile.
The juvenile plumage is rather plain and is easily distinguished from juvenile Temminck's Stint, which has greenish-yellow legs too, but completely different, characteristically scaly upperparts formed by grey feathers with pale, golden-tan fringes.

Note the streaked crown and pale supercilium, which extends to the bill, a feature that distinguishes it from juvenile Long-toed Stint, the other stint with greenish-yellow legs. Furthermore, Long-toed Stint has a very long middle toe and bolder head markings. See this. 8.8. California. DP.

▶ Adult non-breeding.
Least Sandpipers and Long-toed Stints in winter plumage are near identical.

Therefore, it is important to take note of general appearance, the length of the middle toe and the voice.

Least Sandpiper has a small body on short legs and often crouches when feeding. Long-toed Stint is more active, has longer legs and is often seen standing erect and long-necked.

In Least Sandpiper, the bill, tarsus and toes are roughly the same length. In Long-toed Stint the middle toe is longer than the tarsus, on average 5 mm longer than its relative's. 22.12. Mexico. DP

LONG-TOED STINT
CALIDRIS SUBMINUTA

Meaning of the name
'Not quite so small sandpiper'
Calidris from Greek *kalidris* or *skalidris*, a greyish aquatic bird species, mentioned by Aristotle.
Sub, not quite, *minuta*, small.

Jizz
L. 13–16 cm. Ws. 26–31 cm.
One of the seven stints, the only Asian species among the three with yellow legs. In all plumages resembles its American counterpart, Least Sandpiper, but although small, it has a longer neck, legs and toes than this and the other five stints.

The bill is dark and slightly curved with a blunt tip. Perched and feeding birds have a more erect, alert stance than Least Sandpiper, and when disturbed frequently assume a giraffe-like pose with stretched necks.

In flight. Narrow white wing-bar and black longitudinal central bar to the white rump and grey tail. The toes extend beyond the tail, which is not the case for Least Sandpiper.

Similar species
Resembles Least Sandpiper. See this and captions.

Plumage and identification
Adult breeding. Dark bill, sometimes faintly greenish at the base of the lower mandible. The head is heavily streaked in grey and blackish-brown, as are nape, mantle, neck and the sides of the breast. Also, the head has a rusty hue to crown and ear coverts, and the sides of the breast display a tan wash.

The streaked breast is clearly demarcated towards the remainder of the white underside, bringing Pectoral Sandpiper to mind.

The scapulars are dark-centred with rusty fringes and greyish-white tips.

The tertials are dark with wide straight, rusty fringes, not wavy as in Least Sandpiper.
Adult non-breeding and juvenile. See captions.

Voice
The flight call is a short, mellow '*dlloohooh*' and the alarm call is a short, snappy '*dooh-itt*'.

Adult breeding in fresh plumage. Resembles its American relative, Least Sandpiper, in breeding plumage, but its head and mantle are paler and it lacks wavy cinnamon edges to its tertials as well as pale-rimmed upper tail coverts. The pale fringes to scapulars and body feathers on this individual are very fresh, but soon become worn making the bird appear darker with more marked contrasts. The greenish-yellow base to the lower mandible is just visible, a feature also present in juvenile Least Sandpipers. 28.4. Thailand. HS.

Habitat and migration

A very rare vagrant from Asia to Greece, Ireland, Sweden and the United Kingdom.

Frequents lakeshores, muddy coastlines, salt-marsh and marshy localities inland. Winters in eastern India and southern Asia.

Distribution

Breeds sporadically from south-western Siberia eastwards to the Pacific.

▶ Long-toed Stint
Juvenile Long-toed Stint has more distinct markings and shows more contrast than juvenile Least Sandpiper.

Note the almost tiger-striped appearance with a distinct 'V' on its back and the contrast between black-centred scapulars and tertials, with rusty tan and white fringes. Also note the greyish-black wing coverts with greyish-white fringes. Also, the head markings are more distinct with a dark lore that fades towards the eye, and crown streaks extending to base of the bill. Furthermore, the basal part of the lower mandible has a faint greenish-yellow wash.

The very long middle toe together with its claw is about 10 mm long, longer than the tarsus, and on average 5 mm longer than that of the closely related Least Sandpiper, where bill, tarsus and toes are of near equal length. 8.8. China. DP.

▶ Breeding to non breeding.
A very bleached and ragged individual moulting to winter plumage, with scattered rusty tones to its uppermost scapulars.

Note the faintly greenish-yellow base to the lower mandible, which may sometimes be as dark as the remainder of the bill.

In winter plumage closely resembles its relative, Least Sandpiper, but often assumes a more upright stance. Note the clear demarcation between the sides of the breast and the white belly.

See captions for Least Sandpiper for further details. 18.8. China. DP.

WHITE-RUMPED SANDPIPER
CALIDRIS FUSCICOLLIS

Meaning of the name
'The dusky-necked sandpiper'
Calidris from Greek *kalidris* or *skalidris*, a greyish aquatic bird species, mentioned by Aristotle.
  Fuscus, dusky, *collis*, necked.

Jizz
L. 15–18 cm. Ws. 36–38 cm.
A small, elongated sandpiper, smaller than Dunlin and larger than the stints. The bill is dark and slightly curved, paler yellowish-brown at the basal part of the lower mandible. As is typical for long-distance migrants, standing birds display a long flight feather projection beyond the tail, giving a more elongated appearance than the seven stints.
  In flight. Narrow, white wing-bar and a distinctive white rump with black specks, contrasting with a dark back and blackish-grey tail.

Similar species
The only other sandpiper with a similar dark-speckled white rump is adult Curlew Sandpiper in breeding plumage. Juvenile Curlew Sandpiper has a pure white rump. Stilt Sandpiper has a speckled to diffusely barred rump, but is otherwise of entirely different appearance, with much longer legs and bill.

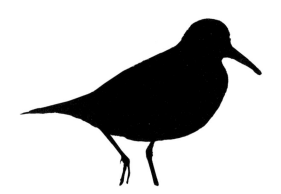

See these. Perched birds are near identical to Baird's Sandpiper. See this and the following captions for similarities and differences.

Plumage and identification
Adult breeding. Sexes are identical. A pale head with streaked, tan crown and ear coverts and a medium-length, dark, slightly curved bill, yellowish-brown at the basal half of the lower mandible.
  The upperparts are greyish-brown with greyish-white fringes and distinct, black feather centres to mantle feathers and scapulars. The uppermost scapulars have a faint rusty hue surrounding the black feather centres. Lesser and median wing coverts are brownish-black with black quills. Neck and breast have triangular blackish-brown markings, expanding to angular markings on the on the sides of the breast and flanks. The remainder of the underside is white. The legs are black.
  Non-breeding. The upperparts are uniform greyish-brown. Breast and flanks are diffusely grey with fine streaks, which in combination with the long flight feather projection distinguish it from the stints.
  Juvenile. See caption.

Adult breeding.
Appears more black-and-white with a cooler tone to its plumage than its similar relative, Baird's Sandpiper, and does not share the tan hue of this species to the feather fringes of upperparts, head, neck and breast.
  The white rump is not visible in perched birds. The best field character is the brownish-black, slightly curved bill with its yellowish-brown base to the lower mandible. Baird's Sandpiper has a black bill.
  Also, notice the long flight feather projection, extending beyond the tertials and tip of the tail. This distinguishes both this species and Baird's Sandpiper from the stints. 8.6. Canada. DP.

Voice
The flight call is a high-pitched, squeaking, crackling '*tziett*', often repeated in a series.

Habitat and migration
A rare, but fairly regular vagrant from North America. Recorded from Austria, Belgium, Czech Republic, Denmark, Finland, France, Germany, Greece, Iceland, Ireland, Italy, the Netherlands, Norway, Poland, Portugal, Spain, Sweden, Switzerland and the United Kingdom.

May be encountered at a variety of costal habitats and inland on wet meadows, and near ponds and lakes. Winters in south-eastern and southern South America.

Distribution
Breeds on moist, coastal tundra with low vegetation in northernmost North America, from Alaska to eastern Canada.

▲▶ Adult breeding.
The white rump with black specks distinguishes this species from all other small sandpipers.
However, note that Curlew Sandpiper in adult breeding plumage has similar markings on its rump and a pure white rump in all other plumages. June. KK.

▶ Juvenile.
The long flight feather projection beyond the tertials and the tip of the tail distinguishes this species from juveniles of the seven stints.
Distinguished from the similar juvenile Baird's Sandpiper by the bill shape and the colour of the lower mandible as well as a generally more greyish, cooler appearance and not least its streaked flanks, which are white in Baird's Sandpiper. See this.
This bird has moulted its first juvenile feathers on mantle and upper scapulars, where they are being replaced by the coming winter plumage. 11.10. The Azores. JSH.

BAIRD'S SANDPIPER
CALIDRIS BAIRDII

Meaning of the name
'Baird's sandpiper'

Calidris from Greek *kalidris* or *skalidris*, a greyish aquatic bird species, mentioned by Aristotle.

Bairdii, named after the American ornithologist, Spencer Fullerton Baird (1823–1887).

Jizz
L. 14–17 cm. Ws. 36–40 cm.

A small, elongated sandpiper, in size somewhere between Dunlin and the stints. The bill is all black and slightly curved towards the tip, with a slimmer tip than than the related White-rumped Sandpiper. This species is a long-distance migrant and on perched birds the long flight feather projection beyond the tail is seen, giving this species a more elongated body shape than that of the seven stints.

In flight. Narrow, white wing-bar and a wide, black, longitudinal bar on the white rump and grey tail. The toes do not extend beyond the tail.

Similar species
Resembles the stints but shows longer flight feather projection. Difficult to distinguish from perched adult and juvenile White-rumped Sandpipers. See these.

Plumage and identification
Adult breeding. Sexes are identical.

The bill is uniformly black with a slender, slightly curved tip.

The crown is dark streaked and the ear coverts faintly tan. The upperpart feathers are brownish-tan with blackish-brown centres and faint rusty markings on mantle and scapulars. The white throat contrasts conspicuously with the streaked neck and breast. Belly and flanks are white.

Black to greenish-black legs.

As a whole, the plumage appears tan, another feature that distinguishes it from White-rumped Sandpiper, which has streaked flanks and a cooler tone to its plumage.

Non-breeding and juvenile. See captions.

Voice
The flight call is a short, soft, warbling trill '*preet*'.

Habitat and migration
A rare North American vagrant. Recorded from Austria, Czech Republic, Denmark, Finland, France, Germany, Greece, Iceland, Ireland, Italy, the Netherlands, Norway, Poland, Portugal, Spain, Sweden and the United Kingdom. May be seen at coastal locali-

Adult breeding.
Size and shape recall White-rumped Sandpiper, with long wings and considerable flight feather projection behind the tertials and tail. However, it differs by having white, unstreaked flanks and a black bill, not bicoloured. Furthermore, in breeding plumage upperparts and breast basically appear tan, not white as in White-rumped Sandpiper. 4. 6. Canada. DP.

ties or inland near lakes and wet meadows.

Winters inland and near coasts of western and southern South America.

Distribution

Breeds on rocky, sparsely vegetated, barren terrain in the Arctic, from eastern Siberia across the northern part of the North American continent.

▶ Juvenile.
Greyish-brown upperparts with distinctly scaly appearance and a faint rusty wash to the uppermost scapulars. Note the distinct black lore flanked by white above and below, which contrasts with the streaked crown and streaked tan breast, clearly demarcated from the white belly and unstreaked flanks. Also, notice the distinctive, long primary projection beyond tertials and the tip of the tail, which distinguishes this species from the stints. 7.9. HS.

▶ Adult non-breeding.
Like its close North American relative, White-rumped Sandpiper, it has greyish-brown upperparts, but has a tan hue to face and breast as well as unstreaked flanks. Also, the bill is black, not bicoloured.

This bird was aged by its large scapulars, so typical for adults, and the few ragged, unmoulted tertials. 28.12. Chile. HJE.

BUFF-BREASTED SANDPIPER
CALIDRIS SUBRUFICOLLIS

Meaning of the name
'The partly red-collared sandpiper'
Calidris from Greek *kalidris* or *skalidris*, a greyish aquatic bird species, mentioned by Aristotle.
　Sub, incomplete, less than, *rufus*, red, *collis*, collar.

Jizz
L. 18–20 cm. Ws. 43–47 cm.
A round-headed and fairly long-necked, benign looking sandpiper with large, black eyes, a short, greyish-black and fine bill and conspicuous orange-yellow legs that nearly match its warm beige, black-speckled plumage.

▲ Juvenile. The combination of uniform greyish-brown upperparts with no pale on the rump is unique among sandpipers. It may be confused with juvenile or adult female Ruffs, but these have longer bills, a characteristic U- or V-shaped white marking around their rumps, distinct white wing-bars. October. USA. KK.

▼ Adult breeding. Adult birds have identical plumage all year round. The male is about 5% larger than the female, making it possible to distinguish sexes at the lek, where males display, standing erect with spread wings accompanied by gutteral, wheezing and rattling sounds. June. Alaska. KK.

In flight. Body the size of a Dunlin, male has distinctly wider and longer wings and its toes are visible beyond the tail. At a glance, similar to a small Ruff or, to a lesser extent, Eurasian Golden Plover.

Plumage and identification
Adult. Plumage is uniform grey-brown all year round with black feathers on crown, nape, sides of the breast and upperparts. The black-centred feathers of the upperparts have wide, grey fringes.

The bill is greyish-black, the legs orange-yellow.
Juvenile. See caption.

Habitat and migration
A regular vagrant from North America to Europe.

Frequents meadows, marshes, stubble fields and embankments near coast and lakes, often together with Ruff and European Golden Plover.

This species is a long-distance migrant, which in autumn presumably migrates first from northern North America to northern South America and from here onwards to eastern South America, an annual round-trip of around 29,000 kilometres.

Distribution
Breeds in Arctic tundra in north-eastern Siberia, Alaska and northern Canada.

Voice
Generally mute but the call is a muffled, short, slightly rasping '*bjoup*'.

▲ Juvenile Ruffs, a Northern Lapwing and in the centre at the bottom, a Buff-breasted Sandpiper. Note the Ruff-like appearance in flight and the telltale differences: Rounded head shape with large black eyes on a beige background, short bill, beige underside and distinct comma markings at the carpal joints. Also, smaller body than female juvenile Ruff, which is somewhere between Buff-breasted Sandpiper and Northern Lapwing in size. 4.10. NLJ.

▼ Juveniles are distinguished from adults by their blackish-brown rather than black feather centres, surrounded by wide whitish fringes, not greyish-brown as in adults.

With its scaly appearance it may resemble juvenile Ruff but may be distinguished by size, head and bill shape as well as leg colour. 22.9. Quessant, France. AA.

SHARP-TAILED SANDPIPER
CALIDRIS ACUMINATA

Meaning of the name
'The pointed sandpiper'
Calidris from Greek *kalidris* or *skalidris*, a greyish aquatic bird species, mentioned by Aristotle.
Acuminatus, pointed.

Jizz
L. 17–22 cm. Ws. 36–43 cm.
The size of a Dunlin and slightly smaller than Pectoral Sandpiper, which it closely resembles. However, its bill is shorter and slightly decurved, the crown rustier and has a distinct white eye-ring. There is a diffuse demarcation between breast and belly.

In flight. All primaries have conspicuously white quills and a white wing-bar runs along the secondary coverts and the three to four innermost secondary coverts. The tail is slightly wedge-shaped, resembling as a hand with fingers pressed together, hence its name.

Similar species
Pectoral Sandpiper, see this.

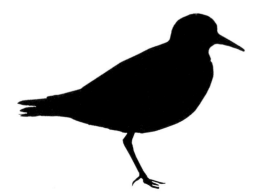

Plumage and identification
Adult breeding. The male is slightly larger than the female and has a longer bill. The crown is dark rusty and black-streaked contrasting with a greyish-white supercilium and a distinct, white eye-ring. Scapulars and tertials have rusty vanes.

The markings of neck and breast distinguish it from Pectoral Sandpiper all year round. They are rusty to beige, heavily streaked and merge into angular markings on belly and flanks. The legs are greenish-yellow.

Non-breeding. Greyish-brown upperparts with large, dark feather centres, bordered by greyish-white vanes and fine dark barring on upper breast and down the flanks.

Juvenile. See caption.

Adult breeding.
The dark rusty crown and distinct white eye-ring give a bolder head pattern than in Pectoral Sandpiper, but first and foremost it is the diffuse transition between breast and belly and the V-shaped marking on the underside in Pectoral that distinguish the two species. 6.5. China. DP.

Voice

The call is a vibrating, short, repeated '*we-veb we-ve-veb*', easy to distinguish from the call of Pectoral Sandpiper.

Habitat and migration

A rare Siberian vagrant, recorded from Austria, Belgium, Bulgaria, Denmark, Finland, France, Germany, Ireland, Kazakhstan, the Netherlands, Norway, Portugal, Sweden and the United Kingdom.

Outside of the breeding season frequents sandy and muddy coastal habitats as well as brackish and freshwater localitities inland.

Winters in southern Asia and further east, from New Guinea across Australia to New Zealand.

Distribution

Breeds on moist, low Arctic tundra in northern to north-eastern central Siberia.

▲▶ Juvenile. Distinguished from adults in breeding plumage by the black feather centres with wide, white fringes and the rufous wash to neck and upper breast, which is only finely streaked.

Distinguished from the similar juvenile Pectoral Sandpiper by rusty scapulars and tertials as well as rusty beige, nearly unstreaked breast. 12.9. Alaska. GV.

▶ Adult breeding to non-breeding, in worn and bleached breeding plumage, but still has a few distinct angular markings on breast and streaked undertail coverts.

Adult non-breeding is very similar to Pectoral Sandpiper, but good field characters are the white eye-ring and the bill, which is shorter and uniformly dark, although sometimes has a slightly paler base to the lower mandible. 20. 8. China. DP.

PECTORAL SANDPIPER
CALIDRIS MELANOTOS

Meaning of the name
'The black-backed sandpiper'
Calidris from Greek *kalidris* or *skalidris*, a greyish aquatic bird species, mentioned by Aristotle.

Greek *melanos*, black, noton, back

Jizz
L. 19–23 cm. Ws. 37–45 cm.
Usually slightly larger and more heavily built than Dunlin, with longer neck. The largest are the size of a small Ruff, the smallest females smaller than Dunlin.

The medium-length, slightly decurved bill is yellowish at the base of the lower mandible, merging into blackish-brown towards the outer third. Seen as a whole, the plumage is nearly uniformly greyish-brown all year round with a diagnostic streaked breast pattern that often tapers to a slight point towards the middle and appears clearly defined agains the pale belly. The legs are greenish-yellow.

In flight. Narrow, white wing-bars on secondary coverts and conspicuously white sides to the rump. The four to six central tail feathers are longer than the rest.

Similar species
Normally easy to identify by its unusual breast pattern, but is very similar to Sharp-tailed Sandpiper. See this.

Plumage and identification
Adult breeding. A wide greyish-white supercilium contrasting with the faintly rusty lore and cheek and the heavily streaked crown. The lower mandible of the bill is yellowish at the base merging into yellowish-brown and black towards the outermost third. The mantle is heavily streaked, and the scapulars have large blackish-brown attenuated feather centres with wide tan borders.

In fresh plumage the uppermost scapulars have rusty rims, creating a pale V-shaped pattern on the back.

Neck and breast are heavily streaked, especially

Adult male breeding.
Within the heavily streaked, voluminous chest is a throat sac, which during the breeding season is used as a sort of bagpipe. The sac is filled with air that when expelled produces a hooting sound during the display flight close to the ground. Some males are sedentary through the breeding season and create leks where females pick their partner. Other males are opportunists that criss-cross the entire breeding area from Alaska to western Siberia, a stretch of around 15,000 kilometres, to seek their fortune in other territories. In neither case does the male participate in incubation or parental care.

Males are not discriminating with partners and several hybrids with other sandpiper species have been described, including with Baird's Sandpiper. 5.6. Canada. DP.

the breast and and the sides of the breast have black arrow-tip markings. The demarcation between the streaked breast and the white belly is abrupt and distinctive. Flanks are diffusely brownish tan and finely streaked. The legs are yellowish.

Female. Smaller than the male with less heavy streaking extending less onto the belly.

Adult non-breeding. Feathers of the upperparts are greyish with light grey fringes. Has greyish upperparts with light grey fringes. In winter plumage, too, the characteristic breast demarcation is distinctive.

Juvenile. See caption.

Voice

The song is an accelerating, mellow '*ooe-ooe-ooe*', performed from low song flight or the ground. The most frequent call is a short, guttural trill, '*preip*'.

Habitat and migration

Pectoral Sandpiper is the most commonly recorded rare vagrant from North America and Asia.

Outside the breeding season it mainly frequents brackish localities, including lakes, fens, wet meadows and coastal lagoons.

Winters primarily in southern South America and to a lesser extent in Australia and New Zealand.

Distribution

Breeds on moist Arctic tundra from Hudson Bay to eastern Canada and across the north of the North American continent to Alaska and from here across Siberia to the Urals.

It is expanding towards the west and may already breed in Europe in small numbers.

▲ Juvenile with slightly bleached plumage, but still with fresh, typical juvenile, wide-fringed feathers with a distinct white V-shape on the back and rusty fringes to the uppermost scapulars. 17.10. Brigantine, New Jersey. HS.

▼ Adult breeding, female.
The female is about 10% smaller than the male and has less extensive breast streaks, appearing more greyish-white. 5.6. Canada. DP.

STILT SANDPIPER
CALIDRIS HIMANTOPUS

Meaning of the name
'The sandpiper with spindly feet'
Calidris from Greek *kalidris* or *skalidris*, a greyish aquatic bird species, mentioned by Aristotle.

Greek *himanto*, leather thongs, and *pous*, foot.

Jizz
L. 18–23 cm. Ws. 38–41 cm.
A very long-legged, long-billed sandpiper, which in body shape and size resembles Curlew Sandpiper, but is easily identified in all plumages by its long, only slightly curved, blunt-tipped bill and its long, greenish-yellow legs.

In flight. Both legs and toes reach far beyond the tip of the tail.

Similar species
May superficially resemble adult breeding Curlew Sandpiper, which has similar markings on its rump and a rusty face.

Plumage and identification
Adult breeding. A diagnostic rusty bar behind each eye contrasts with pale supercilia and a reddish streaked crown. Nape and neck are streaked in blackish-brown, breast and belly are diffusely barred in blackish-brown and faint tan. Mantle and scapulars are dark with black centres decorated with white spots along the sides. Body plumage sometimes shows a rusty hue of varying intensity. Wing coverts are greyish-brown. The legs are greenish-yellow.

In flight. Greyish-brown upperparts with a very narrow wing-bar along primary and secondary coverts, the rump with punctured barring and a grey tail that is white at the centre of most of its outermost feathers.

Female. Less intense markings than males, especially on its breast.

Juvenile. See caption.

Adult breeding.
The name of this species aptly describes the bird, with its long bill and legs.
In breeding plumage, the unique head markings with a combination of a rusty bar behind each eye and rusty sides to the crown. This is presumably a female, with fainter blackish-brown mottled barring than the male, which often displays dense, black mottled barring. 8.6. Canada. DP.

Adult non-breeding. Uniform grey upperparts, neck and flanks are covered in grey mottling and streaks, and the undertail coverts show barring. Apart from its bill, legs and its underside markings, may resemble Curlew Sandpiper in winter plumage.

Voice

Various chattering calls are heard from groups. The most common and characteristic is a short, slightly rasping '*tlloohoohp*'.

Habitat and migration

A rare vagrant from North America. Recorded from Austria, Belgium, Denmark, Finland, France, Iceland, Ireland, Norway, Spain, Sweden and the United Kingdom.

Outside of the breeding season frequents flooded and marshy localities inland, less often along vegetated and muddy coastlines.

Winters from southern North America to central South America.

Distribution

Breeds in North American low Arctic, boggy tundra, near the forest zone.

▲▶ Adult breeding in flight.
May resemble Curlew Sandpiper starting to lose breeding plumage, but lacks red on its underside and also differs by its barred rump, greyish tail with white central bar on the outer tail feathers, and greenish-yellow legs, which protrude far beyond the tail. 7.6. Canada. DP.

▼ Juvenile.
Distinguished from all other sandpipers by its size, the length of its bill and the length and colour of its legs.
Readily distinguished from most *Tringa* sandpipers by the even white edge along lower coverts and tertials. Most Shanks have white notches along this edge. August. USA. KK.

GREAT KNOT
CALIDRIS TENUIROSTRIS

Meaning of the name
'The slender-billed sandpiper'
Calidris from Greek *kalidris* or *skalidris*, a greyish aquatic bird species, mentioned by Aristotle.

Tenuis slender, *rostris*, bill.

Jizz
L. 26–28 cm. Ws. 58 cm.

The largest *Calidris* sandpiper, it is a medium-sized, stocky wader with a medium-length, faintly decurved bill, full breast and medium-length legs that seem too short for the stout body. Similar to Red Knot but slightly larger, more elongated and has a longer bill. Its shape alone renders it unmistakable among other waders.

In flight. Outside of the breeding season has greyish-brown upperparts with very narrow white rump, more or less covered in dark specks, and a grey tail, with the toes not protruding beyond the tip.

Plumage and identification
Adult breeding. Greyish-black, faintly decurved bill and dark legs. The upperparts are greyish-brown with heavily streaked head and black streaks covering breast and flanks, with scattered, black specks on the belly. Most of the large, orange-red scapulars with black centres and white tips are grown during spring migration.

Adult non-breeding. Has uniform greyish-brown upperparts, light to medium grey head and neck, and scattered dark specks on breast and flanks. Legs greyish to greyish-green.

Immature. Breeds at three years of age. It does not assume partial breeding plumage not until the third calendar year.

▲ Breeding to non-breeding.
Breast and flanks are still heavily spotted, but grey scapulars and individual secondary coverts are fresh. In flight, reminiscent of Red Knot, but has a longer, faintly decurved bill and very narrow wing-bars. 25.8. China. 25.8. DP.

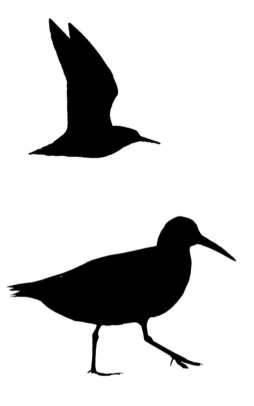

Voice

The call is a short '*voohit*', resembling the call of Red Knot, but shorter and not with a rise in pitch.

Habitat and migration

A very rare vagrant from north-eastern Siberia. Recorded from Germany, Ireland, the Netherlands, Norway, Spain and the United Kingdom. Mainly encountered on coastal mudflats outside of the breeding season.

The most important roosts outside the breeding season, during migration, are located in the northern part of the Yellow Sea. Primary wintering grounds are southern Asia and Australia. Some also winter around the Arabian Peninsula.

Distribution

Breeds on slopes covered in stones and lichen with scattered dwarf vegetaion, in the low Arctic section of the north-eastern Siberian tundra.

▲▶ Juvenile in typical plumage with fresh, well-ordered rows of feathers and wide, pale edges to the feathers. Easily distinguished from juvenile Red Knot by the longer bill, the heavily spotted, blackish-brown breast, and not least the spangled black-and-white upperparts. 28. China. DP.

▶ Adult breeding to non-breeding.
A bird in very worn and faded breeding plumage, but still with dark breast and mantle with a few reddish scapulars among the fresh grey winter feathers. Note that its upper mandible is either very worn or deformed. 23.8. China. DP.

◀ Adult breeding.
The distinct, orange-red scapulars with white tips in fresh plumage are not assumed until spring migration at stop-offs around the Yellow Sea between China and the Korean Peninsula. This bird was photographed at its breeding grounds in Siberia. 5.7. HJE.

Wood Sandpiper.
Adult breeding at the breeding grounds in Finland.
21.6. LG.

The *Tringa* sandpipers (the larger of which are sometimes known as shanks) and their close relatives the Common Sandpiper and Terek Sandpiper, constitute three genera totalling 16 species, which all breed within the Northern Hemisphere.

These species vary from small to medium-sized, and as a group differ from other waders by their elegant proportions, with slim bodies and medium-length neck, bill and legs.

Some species have boldly marked breeding plumages and greyish winter plumages. Others are largely unchanged all year round.

Wary and adroit, these waders often feed while rapidly running to and fro, mostly belly-deep in water, as they hunt for aquatic insects, little fish and amphibians.

Several of the smaller species have a habit of mechanically bobbing their rear body up and down.

In flight the larger *Tringa* species have a falcon-like appearance with angled wings and a pointed falcon-like hand.

The *Tringa* sandpipers may be seen all year round at fresh, brackish or saline localities, along the coast as well as inland. Most species are long-distance migrators, some with summer and winter quarters as far apart as Siberia and Australia.

Unlike *Calidris* sandpipers, these species are generally easy to identify. However, rare American and Asian vagrants have been recorded among European species and these show some great similarities in size and plumage.

SPOTTED REDSHANK
TRINGA ERYTHROPUS

Meaning of the name
'The red-footed wader'
Tringa from Greek *trungas*, thrush-sized, and described by Aristotle as a white-rumped wader that bobs its tail.

Greek *erythropus*, *eruthropous*, red-footed.

Jizz
L. 29–32 cm. Ws. 61–67 cm.
One of the largest *Tringa* sandpipers as well as the most well proprtioned and elegant. It has a long neck, long legs and a distinctive long, bicoloured bill, with a very slight hook or droop at the tip.

Often seen alone or a few birds together. Usually feeds in water to its belly, occasionally swimming and diving.

In flight. Takes to wing with great ease, mostly in a steep curve, nearly always giving its characteristic call. The long straight bill with legs and feet protruding well beyond the barred tail, create a symmetrical flight profile which, together with the white wedge-shaped back, renders it unmistakable.

Similar species
Juvenile and non-breeding may be mistaken for the smaller Common Redshank due to similarities in plumage as well as the colour of bill and legs. See Common Redshank.

Plumage and identification
Adult breeding. Unmistakable with its black plumage, white-speckled upperparts, a white eye-ring and distinctive bill, which all year round is black with contrasting red on the basal half of the lower mandible. Flanks and undertail coverts have varying degrees of white fringes. The female has white specks on her crown and white central undertail coverts.

Legs and feet become black to reddish-brown with or without red areas.

Adult non-breeding. Pale head with grey crown and a well-marked supercilium running from the eye to the base of the bill, contrasting with black lores. The upperparts are pale greyish with whitish fringes and black notches to wing coverts and tertials. The neck and sides of breast are mottled in grey, the belly white, with or without grey-mottled flanks. Legs and feet are red.

Breeding to non-breeding and juvenile. See captions.

Adult, fresh breeding. The characteristic long black-and-red bill distinguishes this species at all times of the year. 27.4. Greece. DP.

Voice

The call is a characteristic, loud and short, sonorous, bisyllabic '*tiooh-it*'.

The song, performed from song flight, begins with the call and continues with a repeated, long, rolling '*chuiiiireeeee*'.

Distribution

On migration Spotted Redshank is seen at marshy coasts and other coastal wetlands, lakes and bogs, often single individuals or small groups. In autumn, however, juveniles gather in larger flocks.

At the wintering grounds this species is found further inland as well, for example at lakes, rice paddies, salinas and marshes.

Breeding biology

The breeding habitat is marshy, wooded fen and tundra near the tree limit. The pair is mostly monogamous, though the female sometimes produces a second brood with a different male. The nest is well concealed, often built within a tussock and lined with plant matter. It is mainly the male that incubates and rears the brood since most females leave the breeding area before the eggs are hatched in early June. Hence the first birds seen on autumn migration are females that after just 2-3 weeks have accomplished their mission and head southwards. The oldest registered ringed bird was at least eight years and seven months old.

▲ Adult breeding, in moulting to non-breeding, with emerging grey feathers and moulting inner primaries. Note the characteristic white wedge-shaped back, which this species has in common with other large shanks, Bar-tailed Godwit and Curlews. 3.7. JL.

▶ Adult breeding, male.
Due to its slender silhouette and black plumage this species is unmistakable among Tringa species and all other European waders. 2.7. Finland. MV.

▲ Juvenile. This bird's legs are folded into the belly feathers, presumably to lessen drag. This is seen in some other long-legged waders too; it alters the flight silhouette, which may confuse the observer.

▲ Adult breeding. Unmistakable in breeding plumage with its black body, red legs and black-and-red bill. Spotted Redshank moults its breeding plumage at staging posts such as the Wadden Sea on its southward migration. This bird is is moulting its flight feathers.

▲ Adult breeding to non-breeding. Both in spring and autumn birds are seen in mottled intermediate plumage, showing elements of summer and winter plumages. 2.8. JL.

Migration

From mid April to mid May, the birds migrate to their breeding grounds in Fennoscandia and Russia. They migrate along western European coastlines and in a wide front across Europe in extended flights, halting at traditional stop-offs like the Wadden Sea, the Evros Delta in Greece and the Black Sea.

Adult females begin their autumn migration as early as June and migration peaks for adult females and males alike in July. Adult birds quickly cross the continent, whereas juveniles follow in August and September and are seen until late October. This species winters sparsely along western coastlines of Europe and around the Mediterranean, but most go to Africa, south of the Sahara.

Distribution

The European population numbers between 20,500 and 54,000 pairs, which breed from northern Scandinavia across northern Russia and nearly to the Berings Strait.

Eastern populations winter from the Arabian Bay across India to south-east Asia.

▲▶ Post juvenile.
The immature bird has an orange marking at the base of the lower mandible and orange legs.
This bird is approaching first winter plumage with an increasingly grey mantle, scattered new scapulars and greater wing coverts, and an increasingly greyish-white belly. 26.8. LG

▶ Adult non-breeding.
Distinguished from juveniles by the nearly uniform white fringes to scapulars and wing coverts, and the uniform grey and white breast and belly, devoid of greyish-brown mottling. 26.8. NLJ.

◀ Juveniles in flight.
Characteristic dusky plumage, streaked and diffusely mottled in greyish-white. Feet and legs are orange-red and the bill is more or less red on its basal half. 10.8. JL.

COMMON REDSHANK
TRINGA TOTANUS

Meaning of the name
'The redshank wader'
Tringa from Greek *trungas*, thrush-sized, and described by Aristotle as a white-rumped wader that bobs its tail.

Totanus, the Italian name for Common Redshank.

Jizz
L. 27–29 cm. Ws. 59–66 cm.
Common Redshank is the most common *Tringa* sandpiper in Europe and is a reference species for the genus.

A medium-sized, greyish-brown, speckled sandpiper with long red legs and a medium-long, bicoloured bill, red at the basal half, black toward the tip.

In flight. The white wedge-shaped back marking

▲ Adult breeding alighting.
Common Redshank in any plumage is unmistakable among waders. No other species has a white wedge-shaped marking on the back, wide white trailing edges to the wings, black-and-red bill and red legs. 20.5. BLC.

▼ Adult breeding.
Common Redshank somewhat resembles Spotted Redshank juvenile and non-breeding. However, can always be identified by its shorter bill, which is red on the entire basal part, not just on the lower mandible as seen in Spotted Redshank. 6.5. LG

and wide white trailing edge to the wing distinguishes this species from any other wader. At the breeding territory the female often alights with quivering wings raised vertically over the back, the white flash of the underwings visible at great distance.

Similar species

Adult may be mistaken for adult Spotted Redshank in winter plumage and juvenile with juvenile Spotted Redshank.

See Spotted Redshank and the photo to the right.

Plumage and identification

Adult breeding. Distinct white eye-ring and bicoloured bill all year round, red at the basal half, the remainder black. Only the basal part of the lower mandible is red in Spotted Redshank.

The upperparts are greyish-brown with dark edges to scapulars and wing covers and the tertials are barred. Neck, breast and the sides of the belly are heavily streaked or diffusely barred. The legs are orange-red to crimson.

Adult non-breeding. Uniform greyish-brown upperparts contrast with paler greyish-brown wing coverts with narrow, pale edges and black streaks. Tertials are either uniform brown or have blackish-brown barring. Neck and breast are more or less diffusely grey and finely streaked. The belly is white with scattered black specks or anglular markings extending to the lower tail coverts. Legs, feet and the basal part of the bill are orange-red.

Other plumages. See photos and captions.

Subspecies

This species is divided into six subspecies, of which the nominate is the most common in Europe.

Apart from this, the darker *T. t. robusta* is common during the winter season. It breeds in Iceland, on the Faeroe Islands and possibly in Scotland.

The remaining four subspecies all breed in Asia.

▲ Common Redshank
Adult Spotted Redshanks and Common Redshank, breeding. A Common Redshank in full breeding plumage in a group of Spotted Redshanks can be difficult to discern, but can be identified easily in flight by the white trailing edge to the wings. Note the shorter bill. 25.4. BLC.

▼ First breeding (2K).
Common Redshank breeds first time when one to two years old. This bird was photographed at its breeding grounds at the Wadden Sea and is aged by the remaining, very ragged, wing coverts and the orange-yellow legs and the faintly red-tinged basal part of the bill. 6.5. LG

▲ Juvenile.
Distinguished from adults by greyish to greyish-red basal part of the bill and the orange-yellow legs and feet. 5.8. JL.

▼ First winter. The nominate form *T. t. totanus*. Aged by remaining juvenile wing coverts and tertials, which still have white fringes and wide black bordering streaks. 9.12. Hyeres. The south of France. AA.

Voice

The call is a characteristic, short and powerful '*doohooh-diooh-diooh-ooh*' in two to four syllables with the stress on the first one.

The alarm call is a repeated, uninflected '*dioohp-dioohp-dioohp*' and the song a more melodious polysyllabic trill, often performed in display flight over its territory.

Distribution

With a population of 340,000 to 484,000 breeding pairs, the Common Redshank is widely distributed in suitable localities across most of Europe. It breeds in marshes, lagoons, fens and damp meadows along the coast and inland. Outside of the breeding season this species is mainly seen along the coastline, and in many places is the most common Tringa sandpiper all year round.

Breeding biology

Breeds at the age of one or two years in monogamous pairs that often return to the same breeding area year after year. The nest with its four eggs is concealed in a straw-domed tussock. The eggs are incubated by both sexes and hatch after about 23 days. Both parents guide their chicks to good feeding areas and adults from nearby territories often aid each other in defence of their young against predators like crows and large gulls. After 10–14 days the male has sole care of the young, which fledge after about 30 days.

The oldest registered Common Redshank was 26 years and 10 months old.

Migration

A very common bird in all seasons and during migration. In spring migrates to and across Europe along coastlines and across a wide front over land during March–May. The breeding period is from April to August depending on latitude. Adults from northern Europe and Fennoscandia move southwards as early as July and in early August most birds are on their way to the winter quarters in southern Europe and north-west Africa. Juveniles are seen on migration until the end of September.

The Icelandic population, of the darker subspecies *T. t. robusta*, mainly winters around the British Isles and the Wadden Sea, but individuals or small groups may be seen along stony and gravelly coasts in western Scandinavia.

North-eastern European birds migrate southwards, passing through Greece on their way to the winter quarters in the Middle East and east Africa.

Distribution

This wader has the widest distribution of all. It is found from Iceland east across Britain, Scandinavia and mainland Europe, continuing across a wide belt through northern Turkey, Kazakhstan, southern Siberia and Mongolia onwards through China to the Pacific.

The wintering quarters are also extensive, stretching from western Europe and west Africa towards the east and south all the way to Australia.

▲ The subspecies *T. t. robusta*, which breeds in Iceland and on the Faeroes, generally has a more brownish-black plumage outside of the breeding season than the nominate.
The breast and belly have a greyish base with heavy markings or streaks merging into mottled barring on flanks and belly. Identified as first winter due to the remaining juvenile wing coverts. 22.1. JL.

▶ As its name indicates, *T. t. robusta*, the Icelandic subspecies of Common Redshank, is more stocky than the nominate, which winters in southern Europe and north-west Africa.
This subspecies may be encountered in surf on stony, ice-filled coasts where it survives on amphipods and other small creatures during winter.
Due to the remaining juvenile wing coverts, this can be identified as a first winter bird. 7.1. LG

WILLET
TRINGA SEMIPALMATA

Meaning of the name
'The half-webbed wader'

Tringa from Greek *trungas*, thrush-sized, and described by Aristotle as a white-rumped wader that bobs its tail.

Semipalmata from *semi*, half, *palmatus*, the palm of the hand. Refers to the partial webbing between the toes of this species.

Jizz
L. 31–41 cm. Ws. 54–71 cm.

The largest and stoutest of the *Tringa* species, with a curlew-like body and a fairly short but heavy, powerful and straight bill.

In flight. Nearly white secondaries and wide white diagonal wing-bar that curves conspicuously on the dark hand. Lacks the white wedge-shaped back of large European *Tringa* species.

▲ Adult breeding. The eastern subspecies, *semipalmata*. The Willet's appearance is transformed in flight, when the otherwise greyish-brown and slightly clumsy-looking bird resembles a huge butterfly with its characteristic black-and-white wing markings. Lacks a white wedge-shaped back marking, and the rump is pure white with no barring evident. 3.4. KK.

▼ Adult breeding, eastern subspecies, *semipalmata*, breeds along the coast of south-eastern USA and the Caribbean. In all plumages it is a little darker with heavier streaking and mottled barring of the upperparts than the western subspecies, *inornata*, which mainly breeds in prairie marshes in western North America. 3.4. Texas. KK.

Plumage and identification

Adult breeding. The bill is fairly short, stout and straight with a grey basal part, gradually merging into black towards the tip. The head and neck are heavily streaked and the upperparts are greyish-brown with curlew-like black barring. Likewise the breast, sides of the belly and flanks show heavy black barred mottling. The remainder of the underside is white and the legs are light grey to greyish-green.

    Non-breeding. The upperparts are pale greyish-brown with narrow, brown quills. The breast has a diffuse greyish-brown wash. The rest of the underside is whitish.

    Juvenile. See caption.

Subspecies

This species is divided into two subspecies: the eastern, nominate form, *semipalmata*, and subspecies *inornata* in the west. The western subspecies is about 15% larger than its eastern counterpart and is generally paler with fewer streaks and less mottled barring. There is some overlap but mostly it is possible to distinguish the two subspecies.

Voice

Several variable calls, but usually a single or repeated, short '*kip*' and a softer, rolling '*kiooh-kiooh-kiooh*'.

Habitat and migration

A very rare vagrant from North America. Recorded from Finland, Norway and France and on several occasions from the Azores. Outside of the breeding season frequents stony, sandy and muddy beaches along coastlines.

Distribution

A North and Central American species with wintering grounds along the coastlines of southern North America, Central America and the northern part of South America.

▲ Adult non-breeding, *inornatus*, has featureless pale greyish-brown upperparts, whereas *semipalmata* shows contrast between greyish-brown scapulars and back feathers and the grey wing coverts. 13.9. Texas. KK.

▼ Juvenile *inornatus* is paler than juvenile *semipalmata*, which has a bolder subterminal band and dark notches along the edges of scapulars and wing coverts. 17.8. KK.

COMMON GREENSHANK
TRINGA NEBULARIA

Meaning of the name

'The wader of the fog'

Tringa from Greek *trungas*, thrush-sized, and described by Aristotle as a white-rumped wader that bobs its tail.

Nebularia, nebula, fog.

Inspired by the ancient Norse name of this species Skoddefoll, fog bird, referring to its habitat, wooded mire, frequently enveloped in fog and mist.

Jizz

L. 30-35 cm. Ws. 68-70 cm.

The largest and most robust greyish-white European *Tringa* sandpiper. Has a long neck, a large head, long greenish-yellow legs and a stout, medium-length, slightly upcurved bill, which is grey on the basal part and black towards the tip.

In flight. A large, stout *Tringa* sandpiper, its wide dark wings contrasting with a light underside with barred underwings. The tail is very pale, faintly barred above. Distinct white wedge-shaped back marking, which extends nearly to the nape and merges with the white of the underside. The toes reach beyond the tail-tip in flight.

Nervous and often wary, takes to the wing at great distance while excitedly repeating its characteristic flight call. Appears more energetic than other *Tringa* species while feeding, usually in small groups running to and fro for fish, sometimes with its whole body emerged in water.

Similar species

Marsh Sandpiper, which is smaller and has a needle-like, slender bill, and the North American vagrant, Greater Yellowlegs, which has yellow legs all year round. See these.

Plumage and identification

Adult breeding. The large head is streaked, as are neck, breast and flanks. The mantle is speckled in black and white, scapulars are greyish-brown and some wing coverts have asymmetrical black centres and streaks.

The Common Greenshank is typically grey and white all year round, but has some heavy black streaking on the mantle and irregular black markings on scapulars and wing coverts. The only similar species among European waders is the Marsh Sandpiper, which resembles a small, dainty Common Greenshank, but has a very slender dark bill. Sweden. 4.7. DP.

The belly and vent are white and the legs are greyish-green.

Other plumages. See captions.

Voice

The call is an easily recognizable, usually trisyllabic, melodious, pulsing whistle '*diew-diew-diew*'.

The song, performed from song flight over its breeding territory, is a unique, rhythmic, throbbing '*piuu-it-piuu-it-piuu-it*'.

▶ First winter.
Note the size, the powerful, upcurved bill, and dark upper wings with no wing-bars. This and the long white wedge-shaped marking on the back, the pale, faintly barred tail and the greyish-green toes protruding beyond the tip of the tail in flight distinguishes Common Greenshank from all other European waders. 9.3. Khok kham. Thailand. HS.

▼ Adult breeding.
Seen on migration on muddy, boggy localities along the coastline and inland alike; any healthy aquatic environment. Usually feeds while walking or running, striking with rapid movements of the bill from side to side. Dragonfly larvae, aquatic beetles, tadpoles, snails, worms, salamanders and small frogs and fish are among its favourites. 14.7. HS.

Distribution

The population numbers 98,700 to 202,000 breeding pairs. This species is commonly seen on spring and autumn migration along coastlines and inland. Birds are often seen individually or in small groups, especially along shallow coastlines, brackish lagoons and river deltas, with sandy or muddy beds with aquatic plants, as well as lakes and bogs.

Breeding biology

Breeds in clearings in wooded bogs and marshes. The pair is mostly monogamous but the male may have an extra partner. The four eggs are laid in a nest on the ground and are incubated by both sexes for 24 days before hatching. When the male has two partners he participates less.

One of the parents leaves the chicks before they are fledged after 25–31 days.

It is unknown at what age birds commence breeding, but one-year-old birds do appear at the breeding grounds.

The oldest registered ringed bird was 24 years and five months old.

▲ Juvenile in flight, aged by its fresh unworn primaries and secondaries and the white fringes along the edges of the secondary underwing coverts. The underwing is has markings similar to most other large *Tringa* species.
Viewed from below, Common Greenshank may resemble juvenile Green Sandpiper, but that species has a barred or mottled breast and pure white underwing coverts. 3.8. JL.

▶ Juvenile. The markings of the upperparts of juveniles show considerable variation. Some individuals, like this one, have greyish-white edges to scapulars and wing coverts and an unbroken, greyish-white edge along the tertials, whereas others have almost chequered, curlew-like upperparts with greyish-white edges, marked with alternating pale and dark notches. Others are intermediate between these two.
However, in late summer, juveniles can always be distinguished from adult birds as the latter have dark specks on the mantle and asymmetrical, scattered, jet-black markings on scapulars and wing coverts. 19.8. LG

Migration

In spring and autumn migrates across a wide front along the coastlines of central and eastern Europe.

Spring migration takes place mainly from mid April to mid May with halts at traditional roosts. Adult females initiate autumn migration from mid June; numbers of both sexes peak in July.

Most juveniles migrate during August and September. Individual birds are recorded as late as the end of October. Winters sparsely along the coastlines of western Europe and the Mediterranean, but its primary wintering grounds are in Africa south of the Sahara.

Eastern populations winter in Africa eastwards from the Arabian Peninsula to India, south-eastern Asia and Australia.

Distribution

Breeds from northern Scotland and northern Fennoscandia eastwards in a wide belt through the Russian and Sibirerian taiga all the way to the Pacific. Some 99% of the European population breeds in Norway, Sweden, Finland and Russia.

▲ First summer with worn, sun-bleached, juvenile flight feathers and ragged, dark brown remnants of juvenile scapulars. The fresh grey feathers are the first signs of adult-like plumage, which will replace the juvenile plumage completely in the course of spring and summer. 28.3. Eilat, Israel. EFH.

▶ Adult non-breeding.
Has a fairly pale, grey-streaked head and uniformly grey upperparts with dark, narrow shafts to greater scapulars and wing coverts and narrow white edges with dark notches on the tertials. Apart from grey flecks on the sides of the breast, the underside is pure white.

The legs are light grey to greyish-green. 17.2. Fuerteventura, Spain. HHL.

MARSH SANDPIPER
TRINGA STAGNATILIS

Meaning of the name
'The wader of stagnant ponds'
Tringa from Greek *trungas*, thrush-sized, and described by Aristotle as a white-rumped wader that bobs its tail.

Stagnatilis from *stagnum*, *stagni*, stagnant water, pond.

Jizz
L. 22–26 cm. Ws. 55–59 cm.
A smaller version of the Common Greenshank, it is a medium-sized, very pale, slender and elegant *Tringa* sandpiper with long, lanky, greenish-yellow legs and a straight, needle-like, greyish-black bill.

In flight. White wedge marking on back. Tail also white but with spots or barring on central feathers. In flight the entire foot projects behind the tail. The underwing coverts are white, not barred or speckled as in Common Greenshank.

Plumage and identification
Adult breeding. The greyish-black bill is needle-like and straight. Head, neck, upper breast and flanks have grey streaks or dark specks. The remainder of the underside may be white or may show varying degrees of black markings. The legs are long and greenish-yellow.

The upperparts are greyish-brown with equally distributed, black feather centres and flank markings.

Similar species
Can only be mistaken for Common Greenshank, which is larger, with stockier proportions and a slightly upcurved bill.

Voice
The call is a short '*chiew*' or '*chiew-chiew-chiew*', quite similar to Common Greenshank's but thinner. The song is a mellow, rhythmically repeated '*dlioohooh-dlioohooh*' and a rhythmical, elastic '*dooip-dooip-dooip*'.

▲ Adult breeding in flight.
Marsh Sandpiper is smaller with daintier proportions than Common Greenshank, which it closely resembles in flight. However, the slender straight bill will always distinguish it from its larger relative. 21.4. HS.

▼ Adult breeding.
The slender bill, long, lanky legs and a generally more elegant appearance make it easy to distinguish Marsh Sandpiper from Common Greenshank. Also, in breeding plumage the Marsh Sandpipern has evenly black-spotted upperparts. 5.6. China. DP.

Distribution

The population of Marsh Sandpiper in Europe totals 12,100 to 30,300 pairs. It has an easterly distribution and is a rare vagrant to western Europe from April to September. Avoids sandy beaches but may be seen at other shallow coasts and inland at lakes, ponds and boggy spots.

Breeding biology

The breeding habitat is swampy terrain in taiga, steppe and brackish marsh. Breeds at one year of age, presumably in monogamous pairs, singly or in loosely knit colonies, usually together with Chlidonias terns, Common Redshanks and Northern Lapwings for better protection of its brood.

The nest is built on an elevated spot in short vegetation, usually close to water. It is lined with dry grass. The four eggs are incubated by both parents and they also share parental care.

Migration

European birds winter in small numbers in the eastern Mediterranean, the majority, however, winter in sub-Saharan Africa from September to March–April. They migrate across a broad front in spring and autumn alike to and from their Russian breeding grounds via the Middle East, the eastern Mediterranean and east of the Black Sea, often in extended legs. It is a scarce migrant in all of western Europe. Asiatic populations winter from the Arabian Peninsula east and south to Australia.

Distribution

The main distribution covers eastern Romania through Ukraine, Russia, Kazakhstan and southern Siberia to northern China. A few breed in Sweden and Finland and there are isolated populations in the Baltics, Poland and Belarus. However, Russia is home to 99% of the total European population.

▲ Juvenile.
As in Greenshank, the plumage of juveniles varies considerably. The whitish vanes along the edge of the greyish-brown feathers of the upperparts may be coarsely serrated or anything in between, as here. The best field character at any season is the slender bill. 19.8. China. DP.

▼ Adult non-breeding.
Has pale greyish upperparts and a nearly white head with a grey crown. Common Greenshank has heavy streaking on crown, neck and the sides of its breast. Marsh Sandpiper and Common Greenshank are often seen together and frequently feed alongside larger birds such as ducks and herons, in order to snatch up prey that has been flushed. 5.2. Thailand. HJE.

GREATER YELLOWLEGS
TRINGA MELANOLEUCA

Meaning of the name
'The black-and-white wader'
Tringa from Greek *trungas*, thrush-sized, and described by Aristotle as a white-rumped wader that bobs its tail.

Greek *melanos*, black, *leukos*, white. Probably refers to this species' spangled, black-and-white breeding plumage.

Jizz
L. 30–35 cm. Ws. 68–70 cm.
The size of a Common Greenshank, which it closely resembles, but also recalls Wood Sandpiper, with finely speckled upperparts and long yellow legs.

In flight. Differs from Common Greenshank by speckled upperparts, the absence of a white wedge-shaped back marking and yellow feet that extend beyond tail. In the smaller Lesser Yellowlegs the feet and the tarsus are visible behind the tail.

▲ Greater Yellowlegs. Adult breeding to non-breeding. Identified by its speckled upperparts, the invariably yellow legs and the lack of a white wedge on the back. Note that secondaries and inner primaries have white notches, which distinguishes it from Lesser Yellowlegs in which they are uniform. This bird is moulting, with unmoulted outer primaries.
2.9. Cape May. TH.

▼ Greater Yellowlegs. Adult breeding.
A graceful *Tringa* sandpiper the size of Common Greenshank, but daintier, with a slightly upcurved bill and the appearance of a Wood Sandpiper. May be distinguished from all European waders by its long yellow legs. Greater and Lesser Yellowlegs both show primary projection beyond the tail; Common Greenshank and Wood Sandpiper do not. 3.4. Texas. KK.

Similar species

Common Greenshank and Lesser Yellowlegs.

Plumage and identification

Adult breeding. Dark, slightly upcurved bill with greyish base. White eye-ring and densely streaked in black on head and neck. The upperparts are greyish-brown, heavily specked in black and white. Bold mottling on breast, belly and flanks extending to the legs.

    Other plumages. See captions.

Voice

The flight call is a short '*diooh-diooh-dioohw*', very similar to that of Common Greenshank.

Habitat and migration

A rare vagrant from North America to a number of European countries. Has been recorded from Belgium, Czech Republic, France, Iceland, Ireland, Italy, the Netherlands, Germany, Norway, Poland, Portugal, Spain, Sweden and the United Kingdom.

Distribution

This North America species is distributed from Alaska through Canada to the Atlantic.

    It winters from southern North America to southernmost South America.

▲▶ Greater and Lesser Yellowlegs, juvenile. This picture demonstrates field characteristics and the size difference and between the two species. Apart from the leg colour, Greater Yellowlegs is similar to Common Greenshank; notice the slightly upcurved bill. Lesser Yellowlegs appears more similar to Wood Sandpiper. The tails of both yellowlegs are finely barred and both lack a white wedge-shaped back marking. 12.9. New York. KK.

▶ Greater Yellowlegs. Non-breeding. Resembles juvenile, lacking the bold black markings on the upperparts and the mottled barring along the flanks that are seen in breeding plumage. 27.1. Florida. KK.

LESSER YELLOWLEGS

TRINGA FLAVIPES

Meaning of the name

'The yellow-footed wader'

Tringa from Greek *trungas*, thrush-sized, and described by Aristotle as a white-rumped wader that bobs its tail.

Flavus, yellow; *pes, pedis*, foot.

Jizz

L. 23–25 cm. Ws. 59–64 cm.

A small, long-necked, elegant, well proportioned Wood Sandpiper-like *Tringa* species with medium-long, dark, slender bill and long, lanky, yellow legs.

In flight. Flight is in typical *Tringa* style but shows no white wedge-shaped back marking. Toes and a little of the tarsus project beyond the tail. In Greater Yellowlegs only the feet reach beyond the tail.

Similar species

Greater Yellowlegs, see this species. Wood Sandpiper, which is a bit smaller, has a shorter bill, and much shorter green to greenish-yellow legs.

Plumage and identification

Adult breeding. Bill is slender and of medium length. Has white eye-ring and heavily streaked head and throat. The upperparts are greyish-brown and heavily speckled in black and white. The under-

Adult breeding.
Near identical to Greater Yellowlegs but not quite as large, with a shorter bill and more elegant proportions. The difference in size may be compared to the difference between Common Greenshank and Marsh Sandpiper. 3.4. Texas. KK.

side is paler than that of Greater Yellowlegs, the breast heavily streaked. There is mottled barring on the flanks; underside otherwise white.

Other plumages. See captions.

Voice

The flight call is a short '*tew*' or '*tew-tew*'.

Very similar to the call of Common Greenshank, but slightly faster, not as lazy-sounding.

Habitat and migration

A rare vagrant to Europe. May be seen at a variety of wetland habitats inland and along the coastline. Recorded from Austria, Belgium, Denmark, Finland, France, Germany, Greece, Hungary, Iceland, Ireland, Italy, the Netherlands, Norway, Poland, Portugal, Slovenia, Spain, Sweden and United Kingdom.

Distribution

This species breeds in boggy clearings in forests and in bogs from Alaska east and south to central Canada. Winters in southern North America, Central America and all of South America.

▲▶ Juvenile. Note that part of the tarsus reaches beyond the barred tail. In Greater Yellowlegs only the feet are visible.

Secondaries and primaries are a uniform greyish-brown with no white notches at the edge of the wing. In Greater Yellowlegs the outer vanes of the secondaries have rows of pale notches. 8.8. KK.

▶ Non-breeding. There is not much seasonal variation in plumage.

The bill attains a yellowish-green hue, and during the winter season the distinct black-and-white of mantle and scapulars is replaced by greyish-brown. The most important characteristics, separating this species from any other *Tringa* sandpiper, are the long yellow legs and fairly long, slender bill. 21.1. Florida. KK.

WOOD SANDPIPER
TRINGA GLAREOLA

Meaning of the name
'The wader of gravel shores'
Tringa from Greek *trungas*, thrush-sized, and described by Aristotle as a white-rumped wader that bobs its tail.
Glarea, gravel. Presumably refers to the species frequenting gravelly lakes and river beds during migration.

Jizz
L. 19–23 cm. Ws. 34–37 cm.
A small elegant *Tringa* sandpiper the size of a Dunlin, but far slimmer and more elegant with a short, slender bill and yellowish-green legs of medium length. The transition between dark upperparts and

▲ Adult breeding, with very fresh coverts and flight feathers. Wood Sandpiper is distinguished in flight from the similar Green Sandpiper by its greyish-brown, heavily mottled upperparts, less white on its rump and tail and, especially, the white vanes to the outer primaries. 29.3. Eilat, Israel. BLC.

▼ Adult breeding. Distinguished from the similar Green Sandpiper by heavy white mottling on upperparts, bicoloured bill and greenish yellow legs. Green Sandpiper has fewer and smaller white spots, a longer bill and greyish to greenish legs. Note that the primaries are aligned with the tip of the tail. 27.4. LG

pale underside is diffuse on the chest, which distinguishes it from other similarly sized *Tringa* species. Often feeds out in the open with rapid movements, occasionally bobbing its body like Common Sandpiper. When flushed, it takes off very fast, often rising to a considerable elevation while emitting its characteristic call.

In flight. Flight is fast, often banking and turning. Dark upperparts contrast with the white rump and barred tail, the feet projecting beyond its tip.

Similar species

Green Sandpiper, Common Sandpiper, juvenile Common Redshank and the rare American migrant, Lesser Yellowlegs. See these.

Plumage and identification

The plumage hardly varies during the year and juveniles, adult breeding and non-breeding birds are similar.

Adult breeding. Brown-streaked crown, white supercilium, white eye-ring and dark brown lore. The bill is greyish-green to yellowish at the base or inner half and black on the outer half. The neck is brown-streaked and the upperparts are dark brown with black and white streaks to scapulars and wing coverts as well as white notches on the greater secondary coverts and the tertials. The front of the breast is mottled, the flanks have mottled barring and the underside is otherwise white. The legs are greenish-yellow to olive.

In flight. The upperparts are dark brown with white spots on back, scapulars and median coverts. The outermost flight feathers have white shafts. The rump is white and the tail has four to five dark brown bars. The underwings are pale, the axillaries and underwing coverts barred in greyish-brown contrasting with medium grey secondaries and primaries. In favourable light, this distinguishes it from its close relative, Green Sandpiper, which has

Adult breeding. Distinguished from Green Sandpiper by its greyish-white underwing with brownish barring. At the breeding grounds its warning call may be heard from a tuft or the top of a spruce, a hectic, shrill, repeated '*kiew-kiew-kiew*', which has been likened to the sound of a blacksmith's hammer striking the anvil. 15.7. JL.

near-black underwings with distinct white wavy lines on the brownish-black axillaries.

Juvenile and plumage details. See illustrations and captions.

Distribution

Seen on migration in spring and autumn at many boggy localities with fresh or brackish water, inland and near the coast, singly or in small, loosely knit groups.

Breeding biology

Breeds in marshes and bogs in deciduous and mixed forest. Starts breeding at one year old, in monogamous pairs.

The four eggs are laid in a concealed nest on the ground, lined with moss and dry plant matter, or sometimes in a deserted nest in a tree. Both parents incubate the single brood, which hatches after 22–23 days. After 10 days the female leaves the breeding area

and the male rears the chicks alone. The brood fledges after 28–30 days.

The oldest registered ringed bird was 11 years and seven months old.

Voice

The alarm call is a series of repeated staccato '*chieb-chieb-chieb*' notes. The call is characteristic, a two- or three-syllable sharp '*chif-if-if*', almost invariably heard when a bird is flushed, or from birds circling wetlands during migration in search for companions.

The song is a warbling '*tiewdl-tiewdl-tiewdl*'. It is performed during song flight and is somewhat reminiscent of the Woodlark's song.

Migration

Northbound migration covers the period from mid March to early June. The first adult females leave the breeding grounds in June and in July they may be seen migrating southwards. The males and

juveniles follow in August–September. Important roosting locations are the Po Valley in northern Italy and the Camargue in southern France. Only a minority winters around the Mediterranean. Most European birds cross the Sahara to tropical Africa, where they winter.

Distribution

The population of 763,000–1,520,000 pairs breeds across Fennoscandia, the Baltics and Central Europe and eastwards in a wide belt through Central Siberia ending at Kamchatka. Extremely rare breeding bird in Scotland.

About 97% of the European population breeds in Sweden, Finland and Russia, with 12%, 41% and 44% in each respectively.

◀ Adult breeding.
Presumably a southbound female . The female leaves the breeding area soon after the brood has been hatched, leaving the male to supervise and defend the chicks until they are fledged.
This is a good example of how pale feathers, less resistant to light and thus faster to weather and wear off, can make a bird in breeding plumage appear dark, having lost its white feather tips and notches. 4.7. LG.

▲▶ Juvenile. Distinguished from adult by its fresh unworn plumage with white notches on wing coverts and tertials. 9.8. LG.

▶ Juvenile. During autumn migration from June to September both Wood Sandpiper and Green Sandpiper may be found together at inland lakes and brackish lagoons along the coastline.
Wood Sandpiper's greenish-yellow legs and greyish-white underwing with greyish-brown barring distinguish it from Green Sandpiper, which has duller legs and a nearly black underwing with distinctive white wavy lines, especially on the axillaries. 6.8. JL.

GREEN SANDPIPER
TRINGA OCHROPUS

Meaning of the name
'The ochre-footed wader'
Tringa from Greek *trungas*, thrush-sized, and described by Aristotle as a white-rumped wader that bobs its tail.

Greek *okhros*, ochre and *pous*, foot.

Jizz
L. 21–24 cm. Ws. 57–61 cm.

A small, elegant sandpiper with a medium-long, greyish-black bill and fairly short, grey to greyish-green legs. It is a bit larger and slightly stockier than the slender Wood Sandpiper. At a distance, gives a characteristic black-and-white impression with blackish-brown, finely white-speckled upperparts contrasting with the white underside. Perched

birds have little or no primary projection, depending on how worn the tail and flight feathers are. A wary bird, it feeds near vegetation, alone or in small groups. When flushed, rises very fast high overhead in a zig-zagging flight, repeating its characteristic warning call.

In flight. White rump, and white tail with only three to four black bars. Seen from a distance, it almost resembles a House Martin. At closer range it reveals characteristic narrow, white wavy lines to the axillaries and white mottled barring to median and lesser coverts on an otherwise brownish-black underwing. Only the tips of the toes extend beyond the tail.

Similar species

Wood Sandpiper is similar, but paler with longer, yellowish-green legs.
See this for details and differences.

▲◀ Juvenile in flight.
The blackish-brown upperparts with small, white specks and predominantly white on rump and tail is characteristic for Green Sandpiper. Note that the outermost primaries are uniformly blackish-brown. In Wood Sandpiper the shafts are white. 24.7. JL.

◀ Adult breeding.
Green Sandpiper may easily be mistaken for Wood Sandpiper due to their similar plumages and leg colors, especially worn, brownish and bleached birds like this one.
    The well-defined border between the streaked upper breast and the white belly, and the unmarked flanks, are reliable field characteristics in all conditions. 15.5. LG.

▲▶ Juvenile.
In flight, narrow white wavy lines on the axillaries are displayed. This characteristic is unique among European Tringa species. 4.8. JL.

Plumage and identification

The plumage varies little in the course of the year and juveniles, adult breeders and non-breeders are near identical.

Adult breeding. Dark brown to blackish-brown upperparts with scattered, white specks. The bill is of medium length and slightly decurved, with a grey to greyish-green inner half, black towards the tip. There is a more or less distinct white streak in front of each eye and a white eye-ring. Head, neck and upper breast are streaked, with a sharp transition to the white underside. Legs are grey to greyish-blue or greenish.

In flight. In well-lit conditions, shows contrast between white-speckled brown body feathers and coverts and the nearly black hand. The rump and tail are white with only two or three distinct black bars. The underside displays a well-defined contrast between the barred tail and the white belly. The nearly black underwings are characteristic, with narrow white wavy lines on the axillaries. The tips of the toes extend beyond the tail.

Juvenile. See captions.

Voice

The alarm call is a shrill '*dooh-eett*'. In the event of threats to the chicks, a rapidly repeated, short '*bieck-bieck-bieck*' is heard, similar to one of a blackbird's warning call.

The flight call is a pointed, whistling '*tlooheet-iet-iet*'. The song, an unmistakable, mechanical '*telooh-ee-te-looh-ee*', is performed either during song flight over its territory or perched at the top of a broken tree.

Distribution

Green Sandpipers migrate over land and individual birds and small groups may be encountered near ditches, channels, ponds, brooks, lakes, bogs and muddy coastal lagoons with vegetation.

Breeding biology

The breeding habitat is wooded bogs and marches with conifers.

At the end of March, while night frost still freezes over the marshes, the characteristic song of the male Green Sandpiper may be heard as the bird flies rapidly, frequently banking, high over its territory.

The pair is monogamous and breeds well away from other pairs. The nest with its four eggs is occasionally on the ground, but usually in the old deserted nest of a thrush, typically four to six metres up in a tree. The brood hatches after 20–23 days of incubation, mostly by the female, who leaves her mate and chicks after about a week, generally in early June, and commences autumn migration. The brood fledges after about 28 days.

The oldest ringed bird was 11 years and six months old.

Migration

Spring and autumn migration is from late February to mid May. The breeding period is from mid April to late June. From early June the first Fennoscandian birds may be seen migrating southwards. Migrating juveniles are most numerous in August–September. A few winter in south-west Europe, but most European birds migrate to tropical Africa.

◀ Adult at its breeding grounds in wooded marshland in Gribskov, northern Seeland. 3.4. LG.

Distribution

The European population consists of between 616,000 and 1,050,000 breeding pairs with the majority in Finland, Sweden, Latvia and Russia. Its distribution stretches from Fennoscandia, patchily through Germany and Poland, eastern Europe (where it is common), and continues in a wide belt through central Asia to eastern Siberia and northern China.

Eastern populations winter from the Arabian Peninsula across India to south-east Asia.

▼ Juvenile.
Distinguished from adults by its fresh plumage with bright white specks and diffuse medium-grey head and neck, devoid of distinct streaking as seen in adults. Note the characteristic blackish-grey underwing with distinct, wavy lines to the axillaries, a feature that distinguishes it from all other European Shanks. 4.7. LG.

▲ Green Sandpiper in flight typically appears black and white with a small compact body. Toes do not extend much beyond the tail (but a little further if the tail is worn).
In spring may be seen migrating early in the morning, when single birds or small groups high above may resemble House Martins due to the contrasts of their plumage. 2 5.8. JL

SOLITARY SANDPIPER
TRINGA SOLITARIA

Meaning of the name
'The solitary wader'
Tringa from Greek *trungas*, thrush-sized, and described by Aristotle as a white-rumped wader that bobs its tail.
Solitaria, solitary.

Jizz
L. 18–21 cm. Ws. 55–59 cm.
This is a North American sister species to the European Green Sandpiper, which is near identical.

A medium-sized *Tringa* sandpiper, slightly smaller and shorter-winged than Green Sandpiper.

The bill is of medium length, and the legs fairly short, shorter than those of Green Sandpiper.

Perched birds show little or no primary projection beyond the tail, depending on age and wear of tail and flight feathers.

In flight. Fast, darting flight, often banking and turning. Seen from below, resembles Green Sandpiper with its white body and near black underwing. Note the characteritic white wavy lines to axillaries and the dark barring to median and lesser coverts on an otherwise brownish-black underwing.

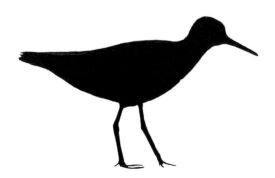

▲ Solitary Sandpiper may be safely identified in flight from Green Sandpiper by its dark rump, and characteristic tail pattern with the tail coverts forming a narrow central stripe, bordered by dark barring that runs the length of the tail.

Green Sandpiper has a white rump and only three to four bars towards the tip of the tail. 30.8. USA. TH.

▼ Adult breeding.
Solitary Sandpiper is very similar to its European sister species, Green Sandpiper. To distinguish perched birds is near impossible, although the Solitary Sandpiper is slightly smaller and its eye-ring is more distinct. The legs are always greenish to greenish-yellow and not grey to greyish-blue as in some Green Sandpipers. 17.5. USA. KK.

The upperparts are dark with little, white specks to back feathers, scapulars and secondary coverts.

The tail is distinctive with its dark uppertail coverts creating a dark longitudinal bar towards the tip, their blackish-brown mottling extending from the trailing edge of the wing to the tip of the tail. This distinguishes it from Green Sandpiper, which has a distinct white rump and only three to four bars towards the tip of its white tail.

Subspecies

Apart from the nominate form, *T. s. solitaria*, one other subspecies is recognized, namely *T. s. cinnamomea*, in which juveniles in particular are reported to have a more cinnamon hue to the upperparts. However, changes from wear and bleaching makes identification of the two subspecies tricky.

Plumage and identification

Adult breeding. Medium-long, slightly decurved bill, greyish-green towards the base, dark towards the tip. Dark lore, a more or less distinct wide, pale streak in front of each eye and a white eye-ring. The upperparts are brownish-black with white specks on mantle and scapulars and white notches on wing coverts and tertials. The heavy streaking of crown and neck extends to the upper breast with a well-defined, clear-cut transition to the white underside. The foremost part of the flanks has mottled barring. The legs are greenish-yellow but not as yellow as in Wood Sandpiper, and not grey to greyish-blue as in some Green Sandpipers.

Juvenile. See caption.

Other plumages. Adults and immatures are very difficult to distinguish in winter and are not described here.

Voice

The call is a short '*tweeet-twiet*', the second syllable higher than the first, higher-pitched and finer than the trisyllabic call of Green Sandpiper.

Habitat and migration

A rare vagrant to Europe from North America. Has been recorded in France, Iceland, Ireland, Portugal, Spain, Sweden and the United Kingdom. Outside of the breeding season, it is more often seen inland near fresh or brackish water than near the coast.

Distribution

Breeds in swampy light forest from Alaska and eastwards in a wide belt across Canada.

Wintering range is from southern North America to central South America.

Juvenile.
Distinguished from adult by its blackish-brown crown and diffusely, blackish-brown breast with no heavy streaking, and in fresh plumage by the distinct white specks and clear-cut notches, particularly on the tertials. 30.8. USA. TH.

COMMON SANDPIPER
ACTITIS HYPOLEUCOS

Meaning of the name
'The white-bellied coast-dweller'
Greek *aktites*, coast-dweller. *Hypoleucos* from *hupo*, beneath, *leukos*, white.

Jizz
L. 19–21 cm. Ws. 38–41 cm.
A small short-necked and short-legged sandpiper with a medium-long bill and a horizontal posture. Constantly bobs its tail-end. On perched birds the long tail is seen to extend well beyond the tips of the folded wing.

In flight. Flight is usually low above water, on arched wings with short, shallow wing-beats alternating with gliding. The wide white wing-bar, unique among small *Tringa*-like sandpipers, is conspicuous.

Similar species
Perched birds may be mistaken for Wood Sandpiper, Green Sandpiper and Spotted Sandpiper, a rare vagrant from North America. See these.

▲ Juvenile in flight.
The plain brown upperside except for white-barred outer tail and the wide white wing-bar, extending to the body, is characteristic. The latter also distinguishes it from its American sister species, the Spotted Sandpiper, in which the wing-bar ends near the middle of the arm. 2.8. NLJ.

▼ Adult breeding.
Resembles both Wood Sandpiper and Temminck's Stint, but may be distinguished from all European sandpipers by the characteristic white spur between its breast and the wing's leading edge, as well as by its long tail. 11.6. JL.

Plumage and identification

Adult breeding. Medium-length, fairly powerful bill, yellowish at the base, greyish-black towards the tip. Pale head with finely streaked crown, distinct white supercilium, white eye-rings and dark lores, which continue behind the eye as a dark eye-stripe. The shafts of the mantle feathers and scapulars are black and when sunlit a vague bronze hue may be seen.

The wing coverts have a dark stripe across, and the tertials are barred. The breast is greyish-brown with fine streaks and pale at the centre, creating a contrast to the white underside. Between the greyish-brown breast and bend of the wing a characteristic white spur or wedge is usually evident. The legs are greyish-green or yellowish.

In flight. Wide white wing-bars extending to the white sides of the body. The outermost feathers of the long rounded tail are predominantly white, the next ones in have black markings, and the central ones are dark.

Juvenile. Resembles adult, but has narrow beige-brown fringes to feathers and no barring on the tertials.

▲▶ Adult breeding.
Tringa species and their close relatives are associated with bogs and mudflats on migration. But at the breeding grounds several species use posts, withered branches and treetops as both song and lookout posts. Runs for cover among shrubs and branches when disturbed and sometimes feeds on insects and other small creatures while climbing among branches. 4.7. Norway. AK.

▶ Juveniles.
The wide, white wing-bar is translucent on primaries and secondaries. This is further enhanced by white greater secondary coverts that merge with white axilliares, creating an underwing pattern unique to this species. 23.7. LG

Voice

The flight call is a high-pitched, excitable '*hee-dee-dee-dee-dee-dee*'.

The song is reminiscent of the call, a shorter, more melodious '*twie-dee-dee-dee*', heard during its extended display flight, which is performed with bat-like, fluttering wingbeats.

Juvenile.
Distinguished from adult by its fresh unworn plumage with buff edges to its feathers, and tertials with dark notches along the edges instead of barring. 23.7. LG

Distribution

One of the most common and widely distributed European waders. A common visitor during autumn migration at all kinds of rivers, channels, lakes, ponds and man-made reservoirs, muddy lakeshores, brooks and brackish lagoons, in particular along gravelly, stony or rocky coastlines.

Breeding biology

Its preferred environment is rocky or gravelly coastlines or shores of rivers and lakes, in open terrain and forests alike. Breeds at the age of one or older. The pair is monogamous and often returns to the same breeding territory. In suitable locations pairs may breed within short distance of each other. The nest with its four eggs is sometimes built far from water, among shrubs and trees. The eggs are incubated by both sexes and hatch after approximately 22 days. Both parents raise the brood for about one week, at which time the female leaves the family. The chicks fledge after 22–28 days and soon after leave the breeding grounds, heading south, together with the male. The oldest registered ringed bird was 14 years and six months old.

Migration

In spring, northbound birds migrate at night over a broad front across land from mid March to late May. At the end of June the first adult females return southwards. Migration of adult birds peaks in August–September when juveniles begin to appear as well. Adult birds move quickly through Europe, whereas

migration of juveniles peaks as late as August and dwindles through September. A few birds are seen until mid October. The majority migrates to tropical Africa, but a few winter in western Europe.

Distribution

The breeding area has its western limits in Spain in the south and the British Isles and Scandinavia in the north, and stretches eastwards in a wide belt through Europe and Asia to the Pacific. The population numbers 223,500 to 1,457,000 breeding pairs with the largest populations in Russia (57%), Finland (13%) and Sweden (10%).

Eastern populations winter from Africa across the Arabian Peninsula to India, southern Asia and onwards south to Australia.

▲▶ Small groups of Common Sandpipers flying in low, darting flight with rapid shallow beats of curved wings are a common sight along rocky coastlines in mid and late summer.
   The white wing-bar is conspicuous from far away and in flight the typical, squeaky '*hi-di-di-di-di-di*' is almost invariably heard. 26.7. HS.

▶ Non-breeding. Uniform greyish-brown upperparts without the characteristic black wedges and mottled barring of the breeding plumage. The bill is uniformly grey and the breast diffusely grey with no black streaks. 26.2. Oman. HJE.

SPOTTED SANDPIPER
ACTITIS MACULARIUS

Meaning of the name
'The spotted coast-dweller'
Greek *aktites*, coast-dweller. Latin *macularius*, spotted.

Jizz
L. 18–20 cm. Ws. 37–40 cm.
An American sister species to the Common Sandpiper, with similar behaviour and flight pattern. Very distinct in breeding plumage but less so in other plumages. See captions.

Voice
The call is a disyllabic '*pooheet-pooheet*', coarser and not as squeaky as that of Common Sandpiper.

Habitat and migration
A rare vagrant from North America to Europe. Like Common Sandpiper, is found near muddy lagoons, lakes and coastlines. Has been recorded in Europe in Austria, Belgium, Bulgaria, Denmark, Finland,

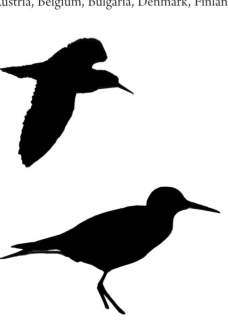

▲ Adult breeding in flight.
Unmistakable in breeding plumage, in which the spotted underside and the orange-yellow bill distinguishes it from Common Sandpiper. Slightly narrower and shorter wing-bar than Common Sandpiper, petering out on the mid primary coverts, whereas it extends all the way to the body in Common Sandpiper and merges with the white sides of the rump. 24.6. Florida. KK.

▼ Adult breeding.
The upperparts have heavier and blacker barring than in Common Sandpiper and the spotted, thrush-like underside is unique to this species during breeding. This is probably a female, having larger and blacker spots than the male. The legs are pink during the breeding season and yellowish in winter. 4.5. Dry Tortugas. Florida. KK.

France, Germany, Greece, Iceland, Ireland, Italy, the Netherlands, Norway, Poland, Portugal, Spain, Sweden, Switzerland and the United Kingdom.

Distribution

A widely distributed North American species with its wintering grounds extending from southern North America to central South America.

▶ Post juvenile.
In fresh plumage, has more distinct tan to whitish fringes to its scapulars and wing coverts.
   Note that the uniform tertials of Spotted Sandpiper have a dark, horizontal subterminal band and are white-tipped, without other markings.
   Juvenile Common Sandpiper has narrow tan fringes along the entire edge of the tertials, interrupted by dark notches. Within the tan tips of the tertials the dark subterminal band extends a little up the edges, creating a U-shaped marking. 11.11. Texas. KK.

▶ Adult breeding to non-breeding.
Predominantly fresh greyish winter plumage on the upperparts but has worn unmoulted tail feathers and bleached, brown flight feathers, and a few unmoulted underside feathers show traces of the black spots of the breeding plumage.
   Spotted Sandpiper usually has a shorter tail than Common Sandpiper, but note that tail lengths vary considerably for juveniles, as well as in adults in worn plumage.
   Note the difference between the juvenile bird at the top and this bird. Spotted Sandpiper should not be identified on tail length alone. The color of the bill and legs, and details of the tertials, must be taken into consideration as well. 17.10. TL.

▶ Adult non-breeding.
A distinct pink hue to the bill, something never seen in Common Sandpiper. All spotted underside feathers of the breeding plumage have been moulted and it has uniformly brown upperparts contrasting with the mottled barring of the coverts. The remaining tertials are worn and devoid of black bars. Also, note the fresh flight feathers just showing beneath the tertials. 31.12. California. DP.

TEREK SANDPIPER
XENUS CINEREUS

Meaning of the name

'The ashen stranger'
Xenus from Greek *xenos*, stranger, and Latin, *cinereus*, ashen.

Jizz

L. 22–25 cm. Ws. 57–59 cm.
An unmistakable medium-sized, short-legged and compact wader with a steep forehead, fairly long upcurved bill and horizontal posture.

Feeds while energetically running about, making rapid head movements and sprints when snatching insects. Often bobs its rear end.

In flight. Similar to Common Sandpiper's flight with shallow wingbeats on slightly arched wings.

Plumage and identification

Adult breeding. The bill is characteristic – upcurved and dark with a yellowish base. Steep forehead, white eye-ring, dark lore and eye-stripe.

▲ Easily identified by its upcurved bill. The back, tail and innermost part of the arm are grey, contrasting to the blackish-brown hand and white trailing edge to the secondaries. 23.3. Gujaret. India. MYJ.

▼ Adult breeding. The upperparts are greyish-brown all year round, but are decorated with black shoulder streaks and distinct black feather shafts during the breeding season. The legs are always orange to yellowish-green. 27.5. HS.

The upperparts are grey to greyish-brown around the year, but in the breeding season show black shoulder streaks and distinct, black shafts to scapulars and wing coverts. The greater secondary coverts and tertials are a darker greyish-brown. The breast has a greyish hue and narrow dark streaks with a well-marked transition to the white underside. The legs are mostly orange-yellow, though in some individuals they are greenish-yellow.

In flight. The body, tail and innermost part of the arm is grey, contrasting with a dark diagonal marking from the lesser primary coverts to the trailing edge of the secondaries.

Other plumages. See captions.

Voice

The flight call is two to five syllables, a rapid '*dooh-dooh-dooh*', not unlike that of Whimbrel but softer.

The song is a characteristic, gull-like '*pe-re-lee*' performed from song flight or perched.

Habitat and migration

A rare but annual visitor to most of western Europe. Is mainly seen on migration over eastern Europe where it breeds. Most frequently recorded in spring near mudflats, sandy coastal ponds and beaches with seaweed. Most winter along the coastline of eastern and southern Africa, across the Arabian Peninsula and southern Asia to Australia.

Distribution

Breeds in lowland forest near rivers and lakes. The European population numbers 15,500 to 50,700 breeding pairs with 98% of these in Russia. Very few pairs breed on the Finnish side of the Bay of Bothnia, a few more in Belarus and Ukraine. From here the breeding area stretches through the Russian taiga and northern Kazakhstan across the central part of Siberia, and almost to the Pacific.

▲ Juveniles are distinguished from adults by the absense of black shoulder streaks, narrower black shafts, and and white-tipped tertials. 27.8. China. DP.

▼ Adult non-breeding. Resembles juvenile, but as in this bird may be distinguished by the contrast between fresh grey winter feathers and worn, brown greater coverts and tertials with worn tips. 27.8. China. DP.

GREY-TAILED TATTLER
TRINGA BREVIPES

Meaning of the name
'The short-footed wader'
Tringa from Greek *trungas* thrush-sized, and described by Aristotle as a white-rumped wader that bobs its tail.
   Brevipes from *brevis*, short, *pedis*, foot.

Jizz
L. 23–27 cm. Ws. 45–54 cm.
Medium-sized among *Tringa* species; slightly smaller than Common Redshank, but more compact with a coarser, medium-length bill and shorter legs.
   Perched birds show no or little primary projection, depending on age and wear. Sometimes bobs its hindquarters.

Similar species
Can only be confused with Wandering Tattler, which in adult breeding plumage has black barring covering its entire underside, undertail coverts included. Juvenile Wandering Tattler has nearly uniform grey upperparts with very narrow, nearly invisible, pale feather fringes and notches; non-breeding birds have a dark grey blotch on each flank.
   Wandering Tattler breeds in north-eastern Siberia and north-western North America and has not yet been recorded in the Western Palaearctic.

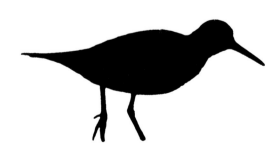

▲ Adult breeding.
Easily distinguished from other *Tringa* species by its squat, robust appearance with a bill of medium length, fairly short, yellow legs and mottled barring across breast and flanks. 22.8. China. HS.

▼ Juvenile.
Distinguished from non-breeding by its uniform light grey upperparts with narrow dark sub-terminal bands, white fringes and notches, most obviously in the greater wing coverts and tertials. 2.10. China. DP.

Plumage and identification

Adult breeding. Thick, medium-sized, greyish-black bill, paler at the basal half. Wide black lore and wide supercilium contrasting with the dark grey crown. The upperparts are greyish-brown but the neck is streaked and contrasts with the mottled barring of breast and flanks. The remainder of the belly is white. The legs are relatively short, sturdy and yellow to greenish-yellow all year round.

Other plumages. See captions.

Voice

A short, melodious '*wieet-wieet*', typical for the genus and not unlike Common Greenshank's call, but more compressed and hectic. It is quite different from the call of Wandering Tattler, which gives a rapid warbling but shrill '*dlee-di-di-di-di*', somewhat reminiscent of the call of Whimbrel.

Habitat and migration

A very rare vagrant to Europe. Has only been recorded from Ireland and Scotland. On migration frequents both muddy and stony environments along the coastline and further inland.

Distribution

Breeds near brooks and rivers in mountainous forest and tundra, and near stony lakeshores.

There are isolated populations in central northern Siberia and north-eastern Siberia. Winters from Taiwan southwards to Australia.

The estimated world population is 40,000–50,000 individuals. It is endangered by hunting and loss of wetlands in its winter quarters due to pollution, drainage and urbanisation.

▲ Non-breeding in flight.
Uniformly greyish-brown upperparts with faint wing-bars along the greater primary and secondary coverts. As seen on the bird at centre bottom, the underwing coverts are dark greyish-brown creating a faint contrast with the paler grey secondaries and primaries. 24.2. Broome. Western Australia. JKAM.

▼ Non-breeding.
Uniformly greyish-brown upperparts and a greyish-brown wash to the breast. The underside and flanks are white.
At close range it shows the nostril grooves extending to the middle of the bill. On the closely related Wandering Tattler the groove extends across the middle of the bill; also juveniles and non-breeding birds have a distinct greyish-brown blotch on their flanks. Sydkorea. AA.

Bar-tailed Godwits. Non-breeding to breeding. With a Eurasian Oystercatcher.

A few of the smaller, shorter-billed males are already red-plumaged and are easily distinguished from the larger and longer-billed females, which are paler and become less intensely reddish in breeding plumage. These birds are the nominate form, *Limosa limosa lapponica*, one of five subspecies. This population winters in western Europe and north-west Africa and constitutes the first group of Bar-tailed Godwits arriving at the Wadden Sea in early spring.
14.4. The Wadden Sea. LG.

There are four species of godwits, all of which breed in the Northern Hemisphere.

Two of these, Black-tailed and Bar-tailed Godwits, are common in Europe, whereas Marbled Godwit and Hudsonian Godwit are North American species, the latter a very rare vagrant to western Europe.

The godwits are large, slender, elegant and well-proportioned waders with long necks, bills and legs, making them easy to spot among other waders.

The breeding plumage is more or less red-toned, and the winter plumage greyish.

The North American subspecies of Bar-tailed Godwit, *L. l. bauri*, holds the world record for long-distance migration, among all animal species. In autumn it migrates directly from its breeding grounds in Alaska to its winter quarters in New Zealand, a distance of some 11,500 kilometres, which it covers in nine days of non-stop flight with an average speed of 60 kph.

BLACK-TAILED GODWIT
LIMOSA LIMOSA

Meaning of the name

'The mud-dweller'

Limosus, muddy. May refer to its frequently muddy bill or a preference for its preference for muddy shores in its winter quarters.

Jizz

L. 36–44 cm. Ws. 70–82 cm.

A large, long-necked and long-legged wader with a very long, straight bill and elegant upright posture. Feeds sedately in shallow water and moist, grassy ground alike, using its long bill to probe and snap at prey. Wary and often noisy at the breeding grounds, where its hoarse, mournful call is heard.

In flight. The flight sihouette is characteristic – long-winged and symmetrical as head and neck project the same distance ahead of the wings as the legs and feet behind. The wide white wing-bar and the white rump are distinctive and in all seasons distinguish it from Bar-tailed Godwit, which has no white wing-bar and only the tips of its toes showing beyond its tail.

Similar species

Bar-tailed Godwit and Hudsonian Godwit, see these.

Plumage and identification

Adult breeding, male. The bill is long and orange-yellow, black towards the tip. Has a pale patch around the eye but not extending beyond into a full supercilium. Black-streaked crown, and reddish-brown head, neck and upper breast, merging into the black barring that covers the greyish-white belly and flanks. Long black legs. The rump is white. The upperparts are greyish with scattered, black-centred, orange-fringed scapulars and back feathers.

Female. Paler with less or no orange tone to head and breast. Slightly larger and longer-billed than the male.

In flight. Distinctive broad white wing-bar, white rump and underwing coverts distinguish it from other godwits.

Non-breeding. The base and inner half of the bill is pinkish, the upperparts greyish-brown and the breast has a grey wash, which contrasts with the white underside.

Juvenile. Greyish bill with pink base. Orange to tan markings on neck and breast. The upper-

Adult breeding. A male with rusty-red neck and breast and distinct black mottled barring to its belly. 14.4. LG.

parts are greyish-brown with yellowish-tan fringes to mantle feathers and scapulars. The lower greater secondary coverts and innermost primary coverts show anchor markings bordered by yellowish-tan. The tertials are black with golden edges and narrow notches.

Subspecies

Apart from the nominate form, *L. l. limosa* that breeds in Europe and Russia, two further subspecies have been described – *L. l. melanuroides*, which breeds in Siberia, and *L. l. islandica*, which is found on the Shetlands, Faeroe Islands, Iceland and Lofoten.

Voice

The flight call is a rapid, mellow '*vooh-vooh-vooh*'

The alarm call is a lamenting, vibrating '*viieew*'. The song is performed from a song flight involving large circles with dipping wingbeats and dives that make the wings produce a wheezing sound. The song often begins with a rapidly repeated '*vi-dooh-vi-dooh*', which transitions into a more elastic, pulsating '*vi-vieetdoohoohe-vi-vieetdoohoohe*'.

Distribution

The breeding habitat is meadows, marshes and soft-soiled damp grassland as well as moors, heaths and steppes with damp depressions.

Outside of the breeding season may be seen in small groups and very large flocks, most often near the coast in well-protected bays and river deltas

▲▶ Adult breeding, female.
The Black-tailed Godwit is unmistakable in flight with its wide, white wing-bar, white rump and white underwing. These three characteristics separate it from all other godwits. 2.6. LG.

▶ Adult breeding. Female.
The female has a longer bill than the male and is usually less red with fewer and fainter black bars on its chest and belly. 15.4. HS.

with tidal mudflats, saline marshes, saline lagoons and sandy beaches as well as in rice paddies inland in Spain and Portugal.

Breeding biology

Breeds at the age of one year or older. The pair is monogamous, staying together for years, and often returns to the same breeding area every year. Here they breed in loosely knit colonies where neighbours participate in joint defence against egg thieves, such as crows and gulls. The nest is built in short vegetation in the open, or partly covered. The female mainly incubates the single clutch, which hatches after 22–24 days.

Both sexes rear the young, which fledge after approximately 30 days.

The oldest known ringed bird was 23 years and seven months old.

Migration

In spring birds migrate along coastlines as well as over land, pausing at traditional roosting and moulting areas.

Adult birds arrive to the breeding grounds from mid March to early June. The breeding period stretches from April to early July. When the chicks are fledged, adults and juveniles gather in larger groups close to the breeding grounds before continuing on their journey in July and August until late October to their winter quarters along the western European coastline and west Africa. Birds from central and eastern Europe migrate, covering a wide front across the Black Sea and the Balkans, towards their winter quarters in Africa from Mali and bordered by Lake Chad towards the east.

Distribution

This species is distributed from western Europe through Russia to the Pacific. The European population is estimated at 102,000 to 149,000 breeding

▲ Juvenile can be distinguished from adult breeding and non-breeding mainly by the lack of mottling on their rufous neck and breast, by the greyish-brown upperside feathers with yellowish-tan rimmed black anchor markings, and by its black tertials with golden edges. 23.7. JL.

▼ Adult non-breeding. Uniform, greyish-brown upperparts with a pink bill, darker towards the tip. Bar-tailed Godwit has a more curlew-like appearance with distinct dark quills, and barred tail and underwings. 10.11. Salalah, Oman. KBJ.

▲ Adult breeding, female. *L. l. islandica.* This subsecies is smaller than the nominate form, but has richer colours and the barring extends all the way to the rump. Iceland. 2.6. DP.

▲ Adult breeding, male. *L. l. islandica.* As in the nominate form, the male's plumage has richer colours than the female's, but with more distinct mottled barring on the underside. Iceland. 13.5. JL

pairs. It is near endangered in most of western Europe due to drainage and cultivation of suitable breeding localities. The largest populations are in the Netherlands with 37% of the population, while Iceland has 20% (*L. l. Islandica*), Russia 17% and Ukraine 11%.

Subspecies

The subspecies *Limosa l. islandica* has a wide distribution with an Icelandic population of 50,000–75,000 individuals, with smaller populations in the Shetlands, the Faeroe Islands and Lofoten (Norway). The majority of birds winter in Ireland and the British Isles and to the Atlantic coastline of France, Spain and Portugal, where both *L. l. islandica* and *L. l. limosa* winter, for example in rice paddies in Extremadura. Others winter in Morocco.

▶ Juvenile, *L. l. islandica.*
The base colour of juvenile *islandica* versus *limosa* is darker reddish-brown, and it has almost sooty upperparts with larger, dark centres to the feathers and wider reddish-yellow fringes. 1.9. The Faeroe Islands. SKKO.

BAR-TAILED GODWIT
LIMOSA LAPPONICA

Meaning of the name
'The mud-dweller from Lapland'
Limosus, muddy. May refer to its frequently muddy bill or a preference for its preference for mudflats in its winter quarters.

Lapponica, from Lapland.

Jizz
L. 37–41 cm. Ws. 70–80 cm.

A large wader, it has a long, slightly upcurved bill, a long neck and medium-length legs. Similar to its near relative the Black-tailed Godwit but is smaller, more compact and less elegant. Is mainly seen in groups, roosting or feeding on sea shores or wetlands. In flight. Flight is rapid and direct on long wings, with toes only just projecting behind the tip of the tail.

Similar species
Perched birds may be mistaken for Black-tailed Godwits, but can be distinguished by their proportions and their upcurved bills. In flight the barred tail and wedge-shaped white back marking distinguishes it from other godwits. In winter plumage the upperparts are greyish-white, mottled and curlew-like, whereas the Black-tailed Godwit has uniform, greyish-brown upperparts. Supercilium is longer than in Black-tailed in all plumages.

Plumage and identification
Adult breeding, male. Bill long and upcurved, sometimes pinkish at its base. Crown streaked with black. The feathers of mantle, back and scapulars have reddish-golden fringes that contrast with the dark-centred, greyish-white wing coverts. Head, neck and the rest of the underside are deep reddish, the legs sooty grey.

Female. Greyish-white with mottled flanks and usually clearly larger and longer-billed than male.

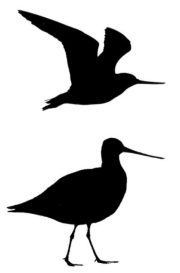

Adults breeding on high-tide sandflats in the Wadden Sea. Females are larger than males on average and differ by their pale plumage and the considerably longer bill. The intensity and extent of the red glow in males signals their strength and health to potential mates, with the most brightly coloured being in the best condition. 4 5. LG.

The bill is pink at the base. Less bright than male; at most a faint rufous wash to head, neck and belly.

In flight. Barred tail, white wedge-shaped back marking and black primary coverts. The underwing has greyish-white coverts with blackish-brown markings, most obviously on the mottled barring of the wide axillaries.

Adult non-breeding. Both sexes resemble females in breeding plumage, with bills that are pink on the basal half.

Juvenile. Very similar to adult non-breeding, but until early autumn can be distinguished by the tan wash to its head, neck and breast. Also note the greyish-brown feathers of mantle and scapulars with tan fringes, contrasting with the paler grey dark-centred wing coverts with wide pale fringes and tertials with coarse, pale sawtooth markings.

Subspecies

Apart from the nominate, which breeds in northern Scandinavia and the northern Russia, this species is separated into four further subspecies with local distribution across Siberia to the Berings Strait and in Alaska.

The two near-identical western subspecies, *L. l. lapponica* and *L. l. taymyrensis*, from the northwestern Siberia, are near-identical. The other three are *L. l. menzbieri* from northern Siberia, which has barring on the tail and a wedge-shaped white back marking and the two eastern subspecies, *L. l. anadyrensis* and *L. l. baueri*, which lack the white back wedge and have heavy mottled barring on the underwing.

The latter three all winter from China and southwards to New Zealand.

The large subspecies *L. l baueri* is the champion among long-distance migrants of the animal kingdom (see page 300).

Distribution

This species breeds only in the northernmost parts of Fennoscandia and in western Russia on heaths and wooded tundra, but is seen commonly on spring and autumn migration along the coastlines of western Europe, the Bothnic Bay and the Baltic Sea. The Wadden Sea is an internationally significant stop-off both in spring and autumn, when thousands of birds may be seen on the mudflats moulting and replenishing their fat reserves. Winters mainly in the Wadden Sea and along the British and Irish coastlines.

Voice

The flight call is a nasal '*kiek-ik*' , also a long series of rapid, warbling notes; '*did-l-did-l-did-l-did-l*'.

The alarm call is a short, shrill '*wuueb*'.

The song, which is performed during song flight, often begins with a few short whistles before progressing to a vibrating '*qu-waeg-qu-waeg-qu-waeg*'.

Adult male breeding, at nest.
The black upperparts with golden streaks conceal the male from predators' eyes when he relieves the female at the nest.
The breeding biology of this species has been poorly studied. Possibly the male is left alone with the brood as in many other waders, while the female heads southwards soon after the eggs are hatched.
5.6. Lapland, Norway. JKAM.

Breeding biology

This species breeds on heaths, wooded tundra and arctic tundra with sparse vegetation. Breeds at two years of age in monogamous pairs. The nest is placed an a dry, elevated spot where the eggs hatch after about 20 days, with the female incubating them at night and the male during the day. Both sexes, or possibly only the male, rear the brood until they fledge at about 28 days, at which time the adults quickly leave the breeding area. The oldest ringed bird was 33 years and one month old.

Migration

On spring migration birds cross western Europe on their way to the Wadden Sea, where they moult to full breeding plumage and replenish their fat reserves before migrating directly to north-western Europe and Siberia. For the birds with the longest routes this is a distance of about 4,500 kilometres.

There are two peaks in spring. The first is during March–April with birds of the nominate subspecies, which winter along the coastlines of north-western Europe and north-west Africa and later at the end of April and first half of May birds of the

▲ It is often possible to determine both age and sex of flying birds, especially in a mixed group like this one.

The left-hand, brightest bird is a male, adult breeding, identified by size, colour and the length of its bill.

The bird at the middle is a juvenile female, identified by the combination of typical buff coloration on breast and belly and the relatively short, not yet fully grown bill.

The right-hand bird is a female, adult breeding, identified by size, colour, flank mottling and its long bill. 19.8. NLJ.

▼ Juvenile. In fresh plumage, has an orange wash to its head, neck and belly, here faded into a faint buff hue. Distinguished from adult non-breeding by the wide pale fringes to the wing coverts and especially the wide blackish-brown tertials with deep, tan sawtooth markings. The mottled curlew-like upperparts furthermore distinguish it from adult non-breeding Black-tailed Godwit, which has uniformly greyish-brown upperparts, and from juvenile Black-tailed Godwit, which has no sawtooth markings on its tertials, has more evenly greyish-brown upperparts and more rufous on its head, neck and breast. JL. 15.9.

subspecies *L. l. taymyrensis* arrive. These winter in west Africa at the coast of Mauretania.

From the peak in mid May and onwards birds of the nominate form migrate towards their breeding grounds in Fennoscandia and north-western Russia, whereas *L. l. taymyrensis* remains at the roosting grounds until the end of of May and beginning of June, when conditions on the Siberian tundra are ideal for nesting.

Adult birds begin return migration in early July, peaking in early August.

Siberian breeding birds, which follow the route across western Europe, only pause briefly before continuing to Africa. European breeding birds stop over for longer in order to moult. From mid August into September juveniles join them. At the end of the year the total number of birds decreases, after most of them have moved on to their traditional wintering grounds in the Wadden Sea, Great Britain and Ireland, with fewer going to other coastlines of western Europe and down to north-west Africa.

Distribution

The European population numbers only 3,700 to 9,000 pairs, all of which breed in Norway, Sweden, Finland and Russia on heaths and wooded tundra. From Russia it is distributed through the northern taiga belt, patchily through Siberia and onwards towards the Bering Strait and far western and northern parts of Alaska.

The wintering area for the nominate form, *L. l. lapponica*, is, as mentioned, along the coastlines of north-western Europe and north Africa.

The subspecies *L. l. taymyrensis* winters along the coasts of east Africa and the Arabian Peninsula to India, with the remaining three subspecies wintering from southern Asia to New Zealand.

▲ First summer.
This species starts breeding at the age of two and most immature birds remain at their wintering grounds until their third calendar year. Some do move a little further northwards, where they moult and then join birds heading for their wintering grounds in autumn.

The bird at the rear is a female and the one at the front a male, identified by length of bills and the partial breeding plumage of the smaller male. Both birds have pale-tipped, fresh secondaries and are moulting their flight feathers. 11.6. LG.

▼ First winter.
Barred tail, white wedge-shaped back marking and a curlew-like grey-and-black appearance of the upperparts. Even at a distance can be distinguished from both European Snipe and Black-tailed Godwit respectively by the long, upcurved bill, and lack of white rump and wing-bars.

Aged by the unmoulted, worn juvenile secondary and primary coverts. 17.1. Oman. HJE.

HUDSONIAN GODWIT
LIMOSA HAEMASTICA

Meaning of the name
'The blood-stained mud-dweller'
Limosus, muddy. May refer to its frequently muddy bill, or a preference for its preference for mudflats in its winter quarters.

Greek *haimatikos*, bloody. Refers to the reddish-brown breast of the breeding plumage.

Jizz
L. 36–42 cm. Ws. 66 cm.
An elegant and well-proportioned North American godwit, reminiscent of a hypothetical blend of Black-tailed and Bar-tailed Godwits with slightly upcurved bill, which is pink near its base all year round.

In flight. As in Black-tailed Godwit, the tail is black, the rump white, but it displays only a narrow white wing-bar. The underwing is distinctive with jet-black axillaries and underwing coverts contrasting with the translucent wing-bar.

Similar species
Black-tailed Godwit and Bar-tailed Godwit. Perched, non-breeding birds display uniformly greyish-brown upperparts and pale belly as in

▲ Adult breeding, male.
The male in breeding plumage is unmistakable with its deep red, diffusely barred underside and undertail coverts and the distinct black triangle of the underwing contrasting with white greater and median coverts. June. Churchill, Manitoba, Canada. KK

▼ Adult breeding.
The female has fainter markings and displays less red than the male. The combination of the wide pale supercilium and the black lore, which is retained in the winter plumage, is not seen in Black-tailed Godwit. June. Churchill, Manitoba, Canada. KK

Black-tailed Godwit, but the bill is slightly upcurved, and the legs are shorter. Bar-tailed Godwit is smaller with distinctly upcurved bill, barred tail and in flight displays a white wedge-shaped back.

Plumage and identification

Adult breeding, male. A wide supercilium under the black crown, and pale face and throat with black streaks. Mantle and scapulars are greyish-black with contrasting white specks. The breast and belly are deep orange-red with distinct black, mottled barring. Undertail coverts are white with wide, black bars. The legs are black.

Female. A little larger than the male and has a longer, more obviously upcurved bill. Its markings are fainter than the male's with red and black mottling on a paler underside.

Other characteristics and plumages: See captions.

Voice

The call is a short, faint '*phiiew*' or '*phiiew-phiee*' at an unvarying pitch. Not as clearly disyllabic as the call of Bar-tailed Godwit, nor nervously vibrating like that of Black-tailed Godwit.

Habitat and migration

A very rare vagrant to Europe. Has been recorded in Denmark, Ireland and Great Britain.

This species migrates for great distances between its breeding grounds in North America and its winter quarters in Argentina and Chile. May travel non-stop for 8,000 kilometres, since no roosting grounds have been recorded along the route.

Distribution

Breeds on moist grassy steppe and heaths near the tree line. Widely distributed in North America from western Alaska to Hudson Bay in Canada.

▲ Juveniles closely resemble juvenile Godwits but may be distinguished by their narrow wingband which does not extend to the body, and the characteristic black triangle on the underwing. All plumages display these characteristics. Notice that the feet protrude beyond the tip of the tail. 3.11. Connecticut. USA. JH.

▼ Adult breeding to non-breeding. In winter plumage may be difficult to tell from Godwit but the remains of mottling from the breeding plumage and the barred undertail coverts are useful clues. The underside of Godwit is uniformaly greyish white. See this and see subspecies, *L. l. islandica.* 6.9. New York. KK.

The three dowitchers all breed on the Northern Hemisphere, with two North American and one Asian species.

Dowitchers are medium-sized, snipe-like waders with snipe-like habits, and godwit-like breeding and winter plumages.

The closely related North American species, both rare vagrants to Europe, are easily distinguished from European waders, but difficult to tell from one another. The differences are in the detail.

Short-billed Dowitcher is separated into three geographical subspecies with some overlap. These are not described, the habitat and migratory route of individual birds sighted in Europe being beyond the scope of this book.

The following field characteristics should be observed when identifying Long-billed and Short-billed Dowitchers, in the following text abbreviated to **LD** and **SD** respectively.

Voice. The calls are distinctive.

LD har a short, shrill '*piit*', often followed by a *Calidris* sandpiper-like, warbling '*pipipipipipi*' when several birds are gathered.

SD has a mellower, often trisyllabic, Tringa sandpiper-like '*tje-de-de*'.

Jizz. Body shape, stature and bill shape.

Females of both species have stockier bodies and longer bills and legs than males. The difference is especially evident in female **LS**, which is easiest to distinguish by its long bill, long legs and the appearance of being very front heavy. At the other end of the scale is male **SD**, easily recognisable by its small, slender, compact body and short bill. Most identification problems arise when distinguishing male **LD** and female **SD**, since they appear near identical.

LD generally has a larger, heavier body, a larger head, thicker neck and longer legs, which are located to the rear of the centre of the body.

A roosting bird compensates for this imbalance by pointing its body diagonally upwards in order to not topple. During feeding, with the bill moving like a sewing machine's needle, it often appears hunch-backed and this, together with its well-rounded belly extending to the fluffy undertail coverts, creates a spherical shape, when viewed from the front and rear as well as side-on. The bill is fairly narrow at its base and straight with a small flat tip. Note that the tip of the bills of large Long-billed females may resemble **SD** but lack the typical bend a quarter of the way from the tip.

SD has a slender body and narrower-looking undertail, and does not give a spherical impession. It has a smaller head, slimmer neck and shorter legs placed at the centre of the body; in roosting birds the stance is horizontal or at a slight angle. From front and rear the bird appears as slender as a *Tringa* sandpiper. The bill is deeper at the base, with a thicker tip and usually shows a characteristic bend a quarter of the way in from the tip, which renders the bill slightly decurved.

Plumage, adult breeding

LD has black-centred feathers with narrow, rusty fringes and the general appearance is of a dark bird with a minimum of rusty colours. Underneath, it is brick-red with black mottling on upper breast and flanks. **SD** has wider, orange fringes to its feathers and and appears paler and more orange. Likewise, the underside is orange-red with black markings on the upper breast and and diffusely barred flanks.

Tail pattern

LD has dark bars, typically wider than the white ones between them.

SD most often has equally wide black and white bars or wider white ones.

Plumage, underwing

LD has pure white lesser coverts on the innermost part of the leading edge of the arm. This creates a distinct white patch.

SD has barring on all underwing coverts.

Juvenile tertials

LD has uniform greyish-brown tertials with narrow white edges.

SD has greyish-black tertials and greater secondary coverts, with distinctive golden fringes and internal tiger stripes.

Plumage, non-breeding

Both species have grey upperparts and pale underside with variable diffuse barring on the greyish flanks.

LD is darker, often with sooty head and neck, giving a grey-hooded impression. The shafts on the upperside feathers are dark, bordered by grey shades, giving it a dark appearance. Often there is a well-defined transition between the grey breast and the pale centre of the belly.

SD has a paler nape and head, paler upperparts with no distinct dark shafts, and a gradual transition to the pale chest.

Note that certain field characteristics, formerly considered safe, have been found to be unhelpful. These include primary projection beyond the tail, the number of visible folded wing tips behind the tertials and, in flight, the length of the toes' projection behind the tail. None of these are reliable, due to wear, moult or too great an overlap between the respective species.

◀ Short-billed Dowitcher. Adult breeding. Identified by its predominantly orange plumage with wide feather fringes to the dark-centred scapulars, the bill's deep base and the subtle bend a quarter of the way in from the tip, which makes its bill appear slightly decurved. Also, this photo was taken in eastern Canada, and Long-billed Dowitcher breeds only in north-western North America. 8.6. Churchill, Canada. DP.

LONG-BILLED DOWITCHER
LIMNODROMUS SCOLOPACEUS

Meaning of the name
'Woodcock-like marsh runner'
Greek *Limnodromus*, from *limne*, marsh and *dromos*, running, a racetrack in ancient Greece.
    Scolopaceus, woodcock-like.

Jizz
L. 24–30 cm. Ws. 46–52 cm.
A medium-sized, snipe-like wader, slightly larger and stockier than Common Snipe. The neck is of medium length, the bill long, as are the legs. These are placed behind the centre of the body, making the bird front-heavy, which it compensates for by tilting the body diagonally upwards when resting. The breast is full and undertail coverts look bulky, which causes it to appear very round around the centre of the body while feeding, crouched like a snipe, the bill moving like a sewing machine's needle.
    In flight. Similar to Bar-tailed Godwit in flight with barred tail and rump, *Tringa* sandpiper-like,

▲ Adult breeding.
The pure white lesser coverts at the innermost part of the arm's leading edge create a distinct, white patch, which is absent in Short-billed, where all coverts on the entire underwing are barred.
    Has longer legs than Short-billed, and often the toes show completely behind the tip of the tail. This may be an extra clue to support identification. 28.4. Salton Sea. California. BS.

▼ Adult breeding.
Distinguished from Short-billed by the narrow, rusty edges and barring of the feathers of the upperparts, and the straight bill with its fine, flattened tip.
    Also told by its larger body, which is held diagonally for balance, to counteract the leg placement (behind centre of gravity). Furthermore, the body is plump behind the legs, giving a rounded shape. 6.4. Texas. KK.

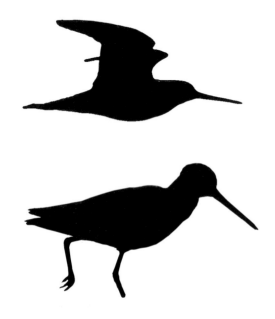

white wedge-shaped marking on back, back and a narrow, white band along the trailing edge of the secondaries.

Plumage and identification

Adult breeding, male. Dark upperparts with narrow, rusty fringes enclosing dark feather centres. Underside brick-red to the tip of the tail, with mottled barring on the flanks and sides of the breast. The legs are yellowish-green. The forehead is generally sloped which results in a flatter head shape than for Short-billed, which is more round-headed with a steep forehead. The bill is greyish-black with the base less deep than in Short-billed, evenly thick, with a fine tip.

Female. Generally larger than the male, heavier body, longer legs and a very long bill. See introduction and captions for further characteristics.

Voice

The call is a distinctive, short, shrill '*pieck*', often followed by a *Calidris* sandpiper-like, warbling '*pipipipipipi*' from birds in group.

Habitat and migration

A rare vagrant from North America. Has been reported from Czech Republic, Denmark, Finland, France, Germany, Greece, Hungary, Iceland, Ireland, Italy, the Netherlands, Norway, Poland, Portugal, Spain, Sweden and the United Kingdom. It is the more nomadic of the two North American species. In its native environment it prefers freshwater locations during migration, but is seen at both coasts and lakes in Europe.

Distribution

Breeds on wet, grassy, Arctic tundra in northeastern Siberia, western and northern Alaska as well as in north-western Canada.

Winters in southern North America and in Central America.

▲ Juveniles of both dowitchers retain their typical feather characteristics until November and are the easiest to identify. Long-billed, as here, has greyish-black scapulars with wide, reddish-brown fringes. The edges of the tertials are narrow. Note that both types of feathers are uniformly dark without internal markings and that the plumage appears evenly patterned, not as in SD, which has a more muddled stripy pattern. 16.10. California. TL.

▼ Adult non-breeding, male. Both dowitchers have grey upperparts and a more or less white underside. Sex and species of this bird is determined by its unbent, short bill, its sooty plumage and the hooded neck and head with its well-marked supercilium and pale crescent below the cheek; also the distinct black, shaded feather shafts on the upperside, and the tail with the black bars wider than white ones. 13.11. Texas. KK.

SHORT-BILLED DOWITCHER
LIMNODROMUS GRISEUS

Meaning of the name

'Grey marsh runner'

Greek *Limnodromus*, from *limne*, marsh and *dromos*, running, a racetrack in ancient Greece.

Latin *griseus*, grey.

Jizz

L. 25–29 cm. Ws. 45–51 cm.

A medium-sized snipe-like wader with a more elegant, *Tringa* sandpiper-like body than in Long-billed, with no suggestion of a hunch-back or heavy rear belly.

The legs are centred on the body, giving the bird a balanced appearance and a horizontal or only slightly inclined stance. Its body is slimmer with a flat back and its central body does not appear spherical while feeding in a stooped posture.

In flight. Similar to Bar-tailed Godwit in flight, with barred tail and rump, *Tringa* sandpiper-like white wedge-shaped back marking, and a narrow pale band along the trailing edge of the secondaries.

▲ Adult breeding.
Identified by the orange coloration not extending over whole underside, the distinctive barred underwing coverts, with no white patch on the lesser secondary coverts at the wing's leading edge, and also the slight bend a quarter of the way from the tip of the bill, seen in most individuals, making bill appear decurved. 31.12. Marco Island, Florida. DP.

▼ Adult breeding.
Identified by the narrow orange rims surrounding black feather centres of scapulars; the orange neck and upper breast; and the bill shape, deep at the base, attenuated towards the tip and ending in a blunt tip. The characteristic bend gives a slightly drooping appearance. 15.5. New Jersey. KK.

Plumage and identification

On the upperparts, feathers are dark-centred with wide orange-yellow fringes. The underside is orange-yellow with speckled sides to the breast and mottled flanks. The legs are yellowish-green. The forehead tends to be fairly steep, giving a rounder-headed appearance than for Long-billed, which has a more sloped forehead and a flatter head shape. The bill is greyish-black, deep at the base with a characteristic bend a quarter of the way from the tip.

Subspecies

There are three subspecies, which are difficult to separate. The prairie form, *L. g. hendersoni* is the only one with a reddish rump in breeding plumage, as Long-billed Dowitcher. The other two have whitish rumps.

Voice

A characteristic, usually bisyllabic and somewhat *Tringa* sandpiper-like '*tiooh-dooh*'.

Habitat and migration

A very rare vagrant to Europe from North America. Prefers saline localities during the winter season.

Recorded from Belgium, France, Germany, Iceland, Ireland, Norway, Portugal, Spain, Sweden and the United Kingdom.

Distribution

Breeds on heaths, bogs with low vegetation and swampy coastal tundra. The three subspecies are geographically seperated and are distributed from south-eastern Alaska across to eastern Canada.

Winters along coastlines from southern North America to northern South America.

▲ Juvenile is distinguished from juvenile Long-billed by its blackish-brown scapulars, greater secondary coverts and tertials, which are gold-rimmed with distinctive internal markings, reminiscent of tiger stripes. The stance is horizontal, typical for SD, and the bill displays the subtle bend near the tip. 4.9. Cape May. TH.

▼ Non-breeding. Identified by its barred secondary wing coverts, with no pure white patch on lesser secondary coverts at the leading edge of the arm. Pale grey head and neck, a deep base to its bill that tapers and bends a quarter of the way from the tip, and fairly short legs. Note the narrow wing-bar on the arm, which is seen in both species. Texas. 15.9. KK.

Curlews at a roost in the Wadden Sea during autumn migration. Close to 8,000 Eurasian Curlews, mostly from Finland and Sweden, stop off here to moult their body and flight feathers - wing moult is apparent in several birds in this image.

Other species present are Shelducks, a few Avocets and, in the green grass, Dunlins. 9.8. L

Curlews are medium-sized to large waders with characteristic downcurved bills and uniformly mottled greyish-brown plumage most of the year.

Eight curlew species are currently recognised, within the genus *Numenius*. Their close relative the Upland Sandpiper is the sole representative of its genus (*Bartramia*). All of them breed on the Northern Hemisphere.

Two of the eight curlew species are probably extinct; the once abundant Eskimo Curlew, *Numenius borealis*, which was last observed in 1981, and the Slender-billed Curlew, *Numenius tenuirostris*, which used to breed in the southern part of western Siberia and may still do so, although no reliable sightings have been recorded since 2001 in Hungary.

This book covers Eurasian Curlew and Whimbrel, which both breed in Europe, Little Curlew and the possibly extinct Slender-billed Curlew, which are rare vagrants from Asia, and Upland Sandpiper, a vagrant from North America.

EURASIAN CURLEW
NUMENIUS ARQUATA

Meaning of the name

'The arched new moon'
Greek, *noumenia*, new moon.
Latin, *arquata*, arched.

Both names refer to the long downcurved bill, which is so characteristic for the genus.

Jizz

L. 50–60 cm. Ws. 80–100 cm.
A large wader with brown streaked plumage, a long slender downcurved bill, long neck and large elongated body on legs of medium length. It is the largest European wader.

Feeds on mudflats or by wading in water to its belly. The bill probes deep, often with prying sideways movements.

In flight. Often seen in small groups with their boat-shaped bodies hanging far beneath long wide wings that are pointed at the tip. Wing-beats are somewhat stiff and gull-like. The toes protrude a little behind the tail, unlike in Whimbrel. A characteristic mellow whistle may often be heard from migrating birds.

Similar species

May be confused with Whimbrel, especially at a distance, but Whimbrel has a shorter bill with a

Adult breeding, female. Sex can be determined by the very long bicoloured bill, which can reach a length of 19 cm and is used to pry into worm and crab tunnels. The finely striped head, and white underwing with no barred mottling on the axillaries, determine the species. 13.4. HS

more marked midway bend (rather than a smooth curve), a dark eye-stripe, blackish crown with a pale central stripe, and striped underwing coverts.

Plumage and identification

Plumage is mottled in brown and white around the year, without much contrast or variation.

Adult breeding. Head, neck and mantle streaked in brown on a beige background. The long down-curved bill is greyish black and reddish at the base.

The scapulars are the darkest part of the plumage, with a wide blackish-brown stripe along the shaft and dark angular markings alternating with white blotches, showing some contrast to the paler wing coverts and tertials, which have a coarsely serrated appearance.

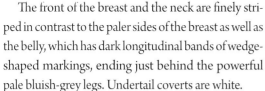

The front of the breast and the neck are finely striped in contrast to the paler sides of the breast as well as the belly, which has dark longitudinal bands of wedge-shaped markings, ending just behind the powerful pale bluish-grey legs. Undertail coverts are white.

In flight, from above. The hand appears decidedly dark with distinct white shafts on the two outermost primaries. The inner primaries, secondaries and lesser coverts are mottled.

The tail is banded, with uppertail coverts mottled in contrast to the conspicuous wedge of white back feathers.

Below. Some individuals have pure white axillaries and underwing coverts. However, most commonly they are faintly striped, some distinctly so.

Female. Similar to the male but larger and with a longer bill.

Juvenile. See caption.

Subspecies

The species is represented by three subspecies.

The nominate form breeds in western Europe, across Scandinavia to the Ural river in Russia. Subspecies *N. a. suschkini* breeds from the Volga and extends eastwards to south-eastern Siberia and northern Kazakhstan, and *N. a. orientalis* breeds from central Siberia eastwards to north-eastern China.

▲◀ Adults breeding.
Distinguished from Whimbrel by the long bill and pale finely striped head with no contrast, as well as by the nearly pure white underwings with no bars on the axillaries. Notice, however, that the barring of the underwing coverts is variable. 6.4. JL.

▲ Adult non-breeding.
Age may be determined by new primaries and secondaries with a few remaining jagged coverts, adult tertials with deep brownish-beige notches, and a new central tail feather. The mottled barring of the tail and white wedge on back are common to Eurasian Curlew and Whimbrel so immature short-billed Eurasian Curlews may easily be confused with Whimbrels. Notice, however, that the tips of the toes protrude beyond the tail, in contrast to those of Whimbrel. 12.10. JL.

Voice

The call is a characteristic, rising, mellow whistle 'cur-lee cur-lee'. On migration often a more extended 'cur-leee cur-leee cur-lee', which is easy to copy and usually causes the bird to answer. At the breeding ground, the warning sound is an angry, repeated 'vy-vy-vy-vooh'.

The song commences with a pair of evocative deep whistles and continues as a rapid bubbling sound. It is usually performed from a song flight with fluttering wing-beats interspersed with gliding.

Habitat

Seen in small groups or individually on migration in spring and autumn along the coast and inland on wet grasslands and marshes. Outside of the breeding season found near mudflats and sand banks in the tidal zone, river deltas as well as on stony and sandy coastal tide pools.

Breeding biology

Breeds on moors, marshes, bogs and wet grassland and increasingly on meadows and pastures. Begins to breed at the age of two and the usually monogamous couple generally returns to the same territory in subsequent years. The nest with its single brood is typically placed in open grassy vegetation. The brood is incubated and protected by both parents and is hatched after 27–29 days. The female leaves the chicks before they are fledged after 32–38 days. The oldest bird registered was at least 31 years and 10 months old.

▲ Juvenile.
Aged by the short bill, little more than half its final length, and the finely striped neck and breast without dark arrow markings or mottling. 10.8. JL.

▼ Adult breeding, male, identified by the short bill. The species is a rare breeder in Denmark on moors and pastures where the male's beautiful bubbling song may be heard over its territory in spring. 13.5. HS.

Migration

Birds arrive at the breeding grounds from March until early May.

Non-breeding birds and adult females initiate autumn migration from mid June and breeding males join in from mid July, followed by juveniles whose migration peaks in mid August. Northern and western European birds primarily winter along the coasts of western and southern Europe but a small proportion of the Irish and British population are sedentary. Eastern populations migrate across the Balkans to winter around the Mediterranean, in west Africa and eastwards to east Africa, the Arabian peninsula and further south and east to southern Asia.

Distribution

The European population numbers 212,000–292,000 pairs, with 33% in Finland, 28% in the United Kingdom and 27% in Russia. The total population has declined significantly in recent decades and the species is classed as Near Threatened due to pressure from habitat degradation as well as destruction of its winter quarters.

Its distribution extends from western Europe eastwards through Russia and central Siberia to north-eastern China.

Adult in breeding plumage and a chick hatched a few days before.

The Eurasian Curlew feeds on a variety of dietary items, depending on season and locality, e.g., worms, crabs, mussels insects, berries and on rare occasions mice, young birds and small fish. 13.5. HS.

WHIMBREL
NUMENIUS PHAEOPUS

Meaning of the name
'The grey-footed new moon'
Greek, *noumenia*, new moon.

Latin, *phaeopus* from Greek *phaios*, grey, and, *pous, podos*, foot.

Jizz
L. 40–46 cm. Ws. 76–89 cm.

Very similar to Eurasian Curlew but distinctly smaller with shorter bill and legs. Has a frowning expression due to its distinct eye-stripe and blackish-brown crown parted by a pale central stripe.

In flight. Has faster and lighter wingbeats than Eurasian Curlew and its toes do not extend beyond the tail. Furthermore, the flight call is easily distinguished from that of Eurasian Curlew.

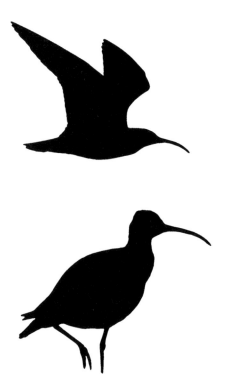

▲ Due to the date, probably a non-breeder. In flight easily distinguished from Eurasian Curlew by its relatively short and very curved bill and by the heavily striped axillaries. The Wadden Sea. 12.6. LG.

▼ Adult breeding during spring migration. Shows intact lighter-coloured notches along the fringes of its feathers and the distinct dark crown with a pale central stripe, typical of this species. 24.4. LG.

Similar species

Eurasian Curlew, especially juveniles that have a relatively short bill. See this species for separation.

Plumage and identification

Plumage is streaked in brown and whitish all year with no great contrast or variation.

Adult breeding. Characteristic head pattern with blackish-brown crown parted by a lighter-coloured central stripe. More or less prominent wide, pale supercilium in front of the eye, framed by the dark lore and eye-stripe. The bill is greyish-black and strongly curved, sometimes pinkish at the base of the lower mandible.

The rest of the head, the neck and the front of the breast are finely striped, extending to the belly where dark, arrow extending to the greyish-blue legs. The flanks appear mottled and undertail feathers are white.

Above, the feathers have blackish-brown centres with lighter-coloured notches along the rim.

In flight. The underwing differs from that of Eurasian Curlew by invariably having distinctly barred axillaries and underwing coverts. The tail and rump is barred, contrasting to the wedge-shaped white back marking. The toes extend beyond the tail.

Juvenile. Distinguished from adults by its fresh plumage.

Subspecies

This species is divided into seven subspecies.

N. p. islandicus is represented by a small population that breeds in north-eastern Greenland and from Iceland southwards over the Faeroe Isles and Shetland to northern Scotland. It winters in western Africa

The nominate breeds from northern Fennoscandia eastwards through Russia to the Jenisej River in Siberia. A scarce winter visitor in south-western Europe, it winters primarily in Africa through the Middle East and further east to India. The two Euro-

▲ Adult. Species identified by the short, strongly curved bill, dark crown and toes that do not extend beyond the tail. Aged by worn scapulars and tertials devoid of paler notches. 8.8. JL.

▼ Adult breeding, *N. p. islandicus*, surprised by a late spell of snow at its breeding grounds in Iceland. Seen from the front, the pale crown stripe rules out any other species. 16.6. Iceland. JL.

pean subspecies are only distinguishable through their measurements. Three Asian and two North American subspecies constitute the remainder.

The Asian *N. p. alboaxillaris*, which breeds on the steppes north of the Caspian Sea, differs by having white axillaries and migrates across the Middle East to its winter quarters in east and south-east Africa.

The two North American subspecies both differ from all others by appearing beige; barred tail coverts and back feathers here replace the white back wedge. There is discussion as to whether one

of these, *N. p. hudsonicus* (Hudsonian Whimbrel), which has been sighted in Europe, including the United Kingdom, may represent a separate species.

Voice

The flight call is an agitated, shrill, staccato '*gooh-gooh-gooh*', very unlike that of Eurasian Curlew.

The song commences with a couple of deep whistles with ascending pitch followed by a rapid, bubbling trill '*doohoohaee goohgoohgoohgooh*'. The warning is a short barking '*gyek-ka*'

Habitat

Primarily breeds in Iceland, Fennoscandia and Russia. A regular spring passage migrant through both coastal and inland Britain.

Individuals and small groups on migration or roosting may be encountered along both rocky and muddy coasts, and further inland near lakes and wetland. During late summer Whimbrels frequent moors and heaths where they feed on berries. Fairly large groups sometimes gather to rest overnight in protected bays and inlets.

Breeding biology

Breeds on moors and swampy woodland, usually at the age of two. The pair is usually monogamous. They build the nest hidden in a depression, then line it with plant matter. Both parents incubate the single brood, which hatches after 22–28 days. Both parents tend the chicks, which are fledged after 35–40 days.

The oldest recorded bird was 16 years and one month old.

Migration

Spring migration is from late March until late May, culminating in early May. It covers a wide front through north-western Europe, across the Mediterranean and eastern Europe. Birds from the large Icelandic population migrate across the western part of the British Isles.

In autumn, migrating birds from Fennoscandia and north-western Russia are seen from late June. These are unsuccessful breeders and presumably the first breeding females. The culmination is late July into August and small groups of roosting birds may be seen until early September. At this time most are already in their winter quarters in western Africa, where most European birds are considered to winter.

Distribution

Breeds from north-eastern Greenland, the Faeroe Islands and Scotland, eastwards across northern Scandinavia and northern Russia, through Siberia to Alaska and Canada.

In Europe there are 343,000–402,000 breeding pairs, 68% of these on Iceland alone. A further 16% breed in Russia and 11% in Finland.

◀ A group of Whimbrels landing on heather-covered heath during autumn migration. Here they forage on nutritious black crowberries. 24.7. KBJ.

▲ Adult after breeding with very worn and ragged breeding plumage. This shows how the white notches at the feathers' fringes wear faster than the darker areas so the bird appears darker at the end of the breeding season, typical for curlews and sandpipers. 23.8. JL.

▼ Juvenile. Juvenile birds can be distinguished from adults; note the fresh feathers with intact white notches along the fringes, most obviously on the tertials, which are typical for the juvenile bird, and by the short bill that is yet to grow to its final length. 4.9. EFH.

SLENDER-BILLED CURLEW
NUMENIUS TENUIROSTRIS

This species is classed as Critically Endangered and is the scarcest wader species of the entire Western Palearctic and south-western Asia. It may already be extinct.

Meaning of the name
'The slender-billed new moon'
Greek *noumenia*, new moon.
Latin *tenuis, tenue*, slender and *rostris*, bill.

Jizz
L. 36–41 cm. Ws. 80–92 cm.
Similar in size to Whimbrel with pale contrasting plumage. The uniformly greyish-black bill is long and slender, very much so towards the tip. The long legs are greyish-black with 'shorts'; pale feathering extending down the tibia. The base colour of the plumage is pure white rather than greyish-white or pale beige as in Eurasian Curlew and Whimbrel.

In flight. Pale upperparts with a white wedge-shaped back and white barred tail, contrasting to the near-black primaries and secondaries. The underwing coverts are pure white, devoid of markings or bars.

The toes extend beyond the tail.

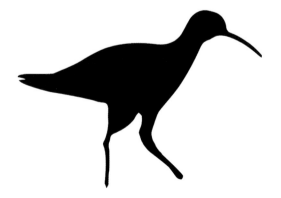

Plumage and identification
Has brown and white upperparts and a white underside with black markings all year round. Adults and juveniles may be distinguished.

Adult. Dark, finely streaked crown with no pale central stripe. The bill is uniformly greyish-black, slender at the base and fine towards the tip.

Detailed studies are the safest way of ruling out Whimbrel and Eurasian Curlew.

The extent of the streaking from head to throat varies significantly from nearly none to fine, dense streaking. The upperparts are blackish-brown with feathers that have dark centres surrounded by lighter fringes. These contrast strongly to the very pale, almost pure white underside, with fine streaks on the front of the breast and black heart or spear shaped markings on the flanks, extending to the very dark, greyish-black legs with feathered tibia.

Non-breeding. Despite the slightly grainy photo, the key features are noticeable. Black-and-white appearance, thin dark bill, dark legs with white 'shorts' and drop-shaped markings that create punctured streaking on breast and belly, all characteristics that distinguish this species both from Whimbrel and Eurasian Curlew. January, 1995. Merja zerga. Morocco. MV.

Mottling on the underside is never seen as it is in the other two other species of Curlews breeding in Europe.

Flight, from above. The tail is white with four or five black bars and, together with the white back wedge and the rest of the upperparts, appears fairly pale in contrast to the nearly black flight feathers and primary coverts. The secondaries are fairly pale with black chequering.

Below. Very pale with conspicuous, pure white axillaries and underwing coverts.

Juvenile. Shorter bill and a beige wash on its breast, which has fine streaking extending to its flanks and none of the distinctive markings seen in adults.

Similar species

Despite variable barring in Eurasian Curlew, especially the subspecies *N. a. orientalis* with its white underwing, all subspecies may be distinguished from Slender-billed Curlew by the combination of size, shape, colour and length of the bill, as well as the markings on the underside and the bluish-grey legs.

Whimbrel is of similar size, but has shorter legs and a more strongly marked head, giving a frowning expression. Juvenile Whimbrels may have nearly un-striped underwings and adults of the subspecies *N. p. alboaxillaris* have pure white axillaries, but all subspecies may be distinguished from Slender-billed Curlew by the shape and colour of the bill, crown and underside markings, generally chequered underwing, leg colour, and toes that do not project beyond the tail in flight.

Voice

The call is mellow, similar to that of Eurasian Curlew 'cur-lee' and may continue as an increasingly high-pitched, bubbling 'plplplplplpliplipliplipli'.

Present status

Despite extensive searches, no documented sightings have been recorded since the early 2000s. Experienced birdwatchers observed four birds in the Evros Delta, Spain in 1999 and there is one more internationally accepted sighting from Hungary in 2001. Additionally, a sighting in 2007 from Uzbekistan of four calling individuals is likely to be reliable, whereas the identity of a photographed individual from Serbia in 2014 is less certain.

It is estimated that the entire world population consists of a maximum of 50 individuals.

Distribution

Presumed to be a scarce breeder in sparsely vegetated bogs at the intersection between woodland and taiga north of the Caspian Sea, between the rivers Volga and Ural in southern Russia and Siberia. Recorded and potential winter quarters are in north Africa and the Persian Gulf.

Non-breeding. The slender dark bill and dark legs, the large markings on breast and flanks and the pure white underwing coverts distinguish this species from Eurasian Curlew and in most cases Whimbrel, but beware with juveniles and the subspecies *N. p. alboaxillaris*. January 1995. Merja Zerga. Morocco. MV.

LITTLE CURLEW
NUMENIUS MINUTUS

Meaning of the name
'The little new moon'
Greek *noumenia*, new moon.
Latin *minutus*, little.

Jizz
L. 28–32 cm. Ws. 68–71 cm.
The size of a Ruff, this is the smallest curlew. The head is small and round with large black eyes and a slender, slightly curved bill of medium length. The long neck and body give a slender appearance. Medium-length legs. The general appearance is of a pale well-proportioned, small curlew.

In flight. At a distance, may be mistaken for a Ruff. The toes extend beyond the tail.

▲ Little Curlew. Adult breeding. Reminiscent of a juvenile Whimbrel with its bill still growing, but differs by its uniform upperparts with no white wedge on back. 25.4. China. DP.

▼ Adult breeding. General appearance is of a very pale bird with pale beige base colour, slimmer and better proportioned than other Curlews. 25.4. Kina. DP.

Similar species

In flight may be mistaken for Ruff or Whimbrel, but may be distinguished from these by the shape of its bill and the uniform upperparts with no white on back and rump. Also, may be mistaken for the rare American vagrant Upland Sandpiper and, less likely, Buff-breasted Sandpiper. See these for details.

Plumage and identification

All year round has blackish-brown, mottled and striped plumage with pale beige as the base colour.

Adult. Dark crown with a very narrow central stripe, a wide, pale supercilium and eye ring encircling large black eyes with a dark stripe behind each.

The bill is relatively short and slender, pink at the base (mostly on the inner half of the lower mandible). The outer half is dark and curved. The neck and front of the breast is finely streaked in brown on a pale beige base. The feathers of the upperparts are blackish-brown and scapulars and wing coverts have large dark centres lined with wide pale fringes. The tertials have dark barring. The belly is pale beige and the legs are pale grey to pale pink.

On perched individuals the wingtip reaches the tip of the long tail, which is dark grey with black barring.

In flight. The upperparts are uniform with no distinguishing marks except the contrast between the dark hand and the pale-fringed wing coverts and back feathers.

The underside is light beige with mottled flanks. The underwing shows evenly striped axillaries and underwing coverts.

Juvenile. Distinguished from adults until around November by their fresh plumage, especially the tertials with small symmetrical white notches along the edges.

Adult breeding.
The short bill, cream axillaries and underwing coverts, very wide and distinct supercilium that blends into the pale neck, the uniform upperparts and toes that protrude behind the tail all distinguish this species from Whimbrel. 7.5. Kina. DP.

Voice

The flight call is a rapid '*qui-qui-qui*', shriller and hoarser than that of Whimbrel.

Habitat and migration

In Europe, this species is a rare vagrant from Asia. It has been sighted in Finland, Germany, Kazakhstan, Norway and the United Kingdom. Sometimes joins groups of European Golden Plovers in autumn on meadows and stubble fields. Appears conspicuously different from these when actively running and walking to and fro during feeding. Rarely seen near coasts, generally near water inland.

Distribution

Breeds in mountainous taiga in central and eastern part of northern Siberia. Winters in Australia.

UPLAND SANDPIPER
BARTRAMIA LONGICAUDA

Meaning of the name
'Bartram's long-tail'
Bartramia after William Bartram (1739–1823), an American botanist, ornithologist and collector. Latin *longus*, long and *cauda*, tail.

Jizz
L. 26–32 cm. Ws. 64–68 cm.
A very long-tailed, medium-sized wader, the size of a Ruff, with a small round head, large eyes, a short straight bill and slender neck.

Similar to European Golden Plover in feeding habits, running, halting and snatching. Sometimes bobs its body up and down like a Common Sandpiper.

In flight. The toes do not protrude beyond the tail.

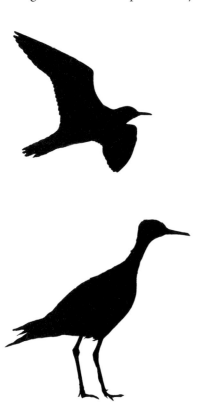

▲ Adult breeding. The wide, heavily barred underwing coverts, the slender body and the short bill is reminiscent of curlews as well as sandpipers. 19.6. Martins Prairie. North Dakota. BS.

▼ Adult breeding. This species is no larger than a Ruff and is often encountered inland on meadows, fields with short grass and other agricultural land where it feeds on insects, worms, small animals and frogs. 5.6. Canada. DP.

Similar species
Little Curlew, less likely Buff-breasted Sandpiper.

Plumage and identification
Appearance similar to a curlew, with mottled and striped plumage year-round.

Adult. Small, round, pale head. Dark crown with a narrow central stripe, large, dark eyes and a short straight bill with dark upper mandible and tip. The neck is streaked, short and slender.

Scapulars are brown with black anchor markings and pale edges that contrast with paler wing coverts. The tertials and the long tail are brown with pale edges and dark barring. The breast is finely strea-ked, the flanks mottled and the underside otherwise greyish-white. The legs are straw-coloured.

In flight. Dark primary coverts contrast with the greyish-brown underside. Pale sides to the rump. The tail, which is rounded when spread, has white tips and a dark sub-terminal band on the outer tail feathers.

The underwing is coarsely barred on axillaries and underwing coverts.

Juvenile. See captions.

Voice
The flight call is a rapid, bubbling '*qui-bib-ib-ib-ib*'.

Habitat and migration
This species is a rare vagrant from North America to Europe. Reported from Denmark, Croatia, France, Germany, Greece, Iceland, Ireland, Italy, Malta, Netherlands, Norway, Portugal, Spain and the United Kingdom.

Found on meadows, fields of short grass and even golf courses, rarely along the coastline.

Distribution
Breeds on prairie, grassland and marginal fields in north-western and central North America. Winters in the central half of South America.

▲ Adult breeding. The long tail and straight bill distinguishes this species from curlews. Notice the sides of the rump, with pale ovals near the base of the tail. 19.6. Martins Prairie. North Dakota. BS.

▼ Juvenile. Distinguished from adult by its fresh plumage with warmer colours and narrow whitish or beige fringes to scapulars and wing coverts, as well as the dark brown tertials. 1.11. Quessant. France. AA.

Snipes are small to medium-sized, brownish and well-camouflaged waders with long bills and relatively short legs. Most species are shy and hard to observe, and many are similar in appearance and difficult to distinguish.

Details such as size, bill length, head and body markings, flight pattern and voice when flushed are all supportive elements in identification, but the most reliable are the differences in their unusual territorial song flights, which, unfortunately, are usually heard only at the breeding grounds

All species except Great Snipe have a rollercoaster-like song flight, with steep dives during which the birds either sing or spread their tail feathers, allowing air pressure to produce a humming or whining 'song'.

The number and width of the spread tail feathers used for this 'song', and the sound they produce, are diagnostic for each species.

Snipes are among the few waders that actively feed their young.

There are 18 species of snipes.

Nine species breed on the Northern Hemisphere, four of these in Europe.

A further two species are sighted as rare vagrants.

The four breeding species are Great Snipe, Common Snipe, Jack Snipe and Pin-tailed or Pintail Snipe.

The remaining two species, Wilson's Snipe and Swinhoe's Snipe, are rare visitors from North America and Asia respectively.

Wilson's Snipe is a 'twin species' to the Common Snipe, a recent split, and Swinhoe's Snipe is an Asian 'twin species' to the Pin-tailed Snipe.

Swinhoe's Snipe will not be described here, since it is near-identical to Pin-tailed Snipe and is best identified by its mechanical song and the width of its outer, stiff tail feathers, which may be seen during song flight over its territory.

Great Snipe, adult breeding.
Singing male at the lek. Courting males inflate their breast feathers and perform their peculiar courtship display, giving a bubbling song and holding the white-tipped tail erect and spread. 15.6. Sverige. DP.

GREAT SNIPE
GALLINAGO MEDIA

Meaning of the name

'The medium-sized hen-like bird'
Gallina, hen, and *ago*, like.

Presumably, the name of the genus refers to the hen-like plumage.

Media, derived from *medius*, medium.

The Great Snipe was formerly placed with the woodcocks in the genus Scolopax and described as having a size between Woodcock and Common Snipe.

Jizz

L. 27–29 cm. Ws. 47–50 cm.
Bill approximately 6.5 cm.

A medium-sized wader and a large snipe with a powerful chest and woodcock-like appearance due to its large eyes and stouter, slightly shorter bill compared to the Common Snipe's.

The body, on the other hand, is slightly larger and bulkier than Common Snipe's, on powerful legs.

▲ Adult breeding at a lek. The white bands of the wing coverts and the heavily mottled flanks, together with the slightly pink, bicoloured bill and the pale legs, distinguish this species from the Common Snipe. That species has greenish-yellow bill and legs, and lacks the row of white bands on the wing. 9.6. Norway. DP.

▼ Adult breeding. The marbled arm with its pale-tipped secondary coverts as well as the distinct white bands along primary and secondary coverts clearly distinguishes this species from Common Snipe. The Common Snipe possesses a single narrow white wing-bar and a wide white trailing edge to the secondaries. 13.6. Sweden. HS.

In flight. Usually flies off with a rumbling and a low 'sneezing' sound when flushed, flashing a brief glimpse of its white outer tail feathers. Flies low and in a straight line for a short distance before diving into shelter after braking abruptly with its tail spread, displaying the wide white markings of the outer tail feathers. In flight the bill is held horizontally or with a slight dip.

Toes protrude beyond the tip of the tail.

Similar species

Common Snipe, see for details.

Plumage and identification

The sexes are identical, as is their appearance throughout the year. Only juveniles may be distinguished.

Overview. Dark crown with a narrow white central stripe, pale supercilium and dark lore with the colour continuing behind the large eyes into the back of the neck, framing a more or less prominent patch below the eye. The bill is dull brownish-pink, deep at the base, and becomes dark towards the tip.

Upperparts are blackish-brown with the pale barbs of the greater scapulars creating wide longitudinal streaks. Buff marbling on scapulars and wing coverts.

Furthermore, lowermost tertial, secondary and primary wing coverts have distinct white tips, which create three pale bands that are especially conspicuous on perched birds.

The breast and belly are greyish-white with a buff flush to the sides of the breast and flanks and dark angular markings on breast and flanks.

The feet are greyish with a faint pink hue.

In flight, from above. The dark upperside displays a distinct, pale wing-bar along the greater wing coverts and two more, constituted by the secondary and tertiary wing coverts. The secondaries – depending on wear – have pale tips, which may appear as a very narrow light band. The four central tail feathers (of usually 16 altogether) are reddish-brown with narrow white tips, whereas the outer ones are nearly pure white. This is best observed during takeoff and when landing. From below. The underwing coverts have wide barring and seem to merge with the diffusely barred flanks.

Juvenile. See caption.

▲ Adult breeding.
The large eye, long bill, heavy body, the trailing edge of the wing with no or only a marginal white trailing edge, and the diffuse barring of the underwing are typical of the Great Snipe. 9.6. Norway. HS.

▼ Adult males at a lek.
Among snipes, only the Great Snipe has a lek where males gather on summer nights to charm the females with their clicking, bubbling song, posturing with tail erect to flash their white tail feathers. 9.6. Norge. HS.

Voice

The song, which is performed at the lek, consists of a series of accelerating clicking sounds, reminiscent of small table-tennis balls rapidly bouncing on a hard surface.

The song is concluded with a fine, vibrating, bubbling call.

Habitat

Has a Fennoscandian and eastern European distribution. Sightings are uncommon during spring and autumn migration, but the species is probably more numerous than it seems, given its discreet life style. It feeds on drier soils than the Common Snipe and is rarely seen out in the open. Furthermore, it sits tight when disturbed and only actively searching suitable localities is it likely to be seen.

Breeding biology

Breeds on high-altitude moors and in damp meadows and bogs in Fennoscandia. The species is polygamous and the only European snipe not to have a song flight. Instead, the song is performed on a lek with short vegetation where males gather at dusk and defend their little knoll. From here they perform their bubbling and clicking song with protruding breast and spread white tail feathers, and perform acrobatic hops, hoping to attract a female.

Adult breeding, at the lek.
The leks of the Great Snipe are found at remote, slightly boggy mountain valleys where the faint bubbling and clicking sounds of the male may be heard in the light summer night. The eyes of this species are the largest among snipes, probably due to this nocturnal behaviour at the breeding grounds. 9.6. Norway. HS.

The nest is lined with plant material and well hidden in vegetation. The female incubates and rears the brood alone. The brood is hatched after 22–24 days and fledged after 21–28 days.

Migration

Migration to and from the breeding grounds covers a wide front across Europe. The species is rarely sighted during spring migration, from early April to early May. In autumn, from early July to early November, it is more plentiful, probably due to the addition of juvenile birds.

In a Swedish research project three birds were fitted with light loggers in order to reveal more about migratory routes and wintering quarters. They all flew from Jämtland in central Sweden directly to Nigeria in just three days, a distance of 6,500 kilometres. Their top speed was 90 kph and the average speed was 70 kph. After a short rest they continued to the Congo Basin, which is assumed to be the wintering area for this species.

Other birds migrate through Europe in smaller steps to their winter quarters in Africa. Small numbers of birds winter in north-western Europe and southern Scandinavia.

Distribution

The world population is 62,500 to 145,000 pairs, which breed from Poland and Fennoscandia eastwards through Russia and northern Kazakhstan to central Siberia. Some 86% breed in Russia, 9% in Norway and 6% in Belarus.

This species winters primarily in Africa, less frequently in the Middle East.

Juvenile. Distinguished from the adult by the whitish fringes to the outer web of the scapulars and the light beige tips of the wing coverts, as well as the 'tiger-striped' tertials and yellowish-green bill and legs. Similar to Common Snipe, but differs by having larger eyes, a shorter bill with a deeper base, rows of white-tipped wing coverts and heavy angular markings on the axillaries, and barred underwing coverts on a grey base. 27.7. JL.

COMMON SNIPE
GALLINAGO GALLINAGO

Meaning of the name
'The hen-like bird'
Gallina, hen, and *ago*, like.

Presumably, the name of the genus refers to the hen-like plumage.

Jizz
L. 25–27 cm. Ws. 44–47 cm. Bill approximately 7 cm.

A medium-sized wader and snipe with a long straight bill that seems extremely long for a bird this size. The breast is prominent and the bird often has a crouched stance when walking on its fairly short legs. Remains hidden in the vegetation most of the day, but may be seen morning and evening feeding along shore vegetation with fast sewing machine-like movements.

In flight. The extremely long bill combined with the snipe jizz makes the flight silhouette easily recognisable. Often takes off at good distance with a hoarse croak when approached and usually rises fast and steeply in rapid zig-zag flight.

▶ Inset is an enlarged section of adult wing coverts from the adult below.
The pale streaks on the upper row of median coverts are parted by a blackish-brown stripe extending to their tips. This is not seen in juveniles where the markings are either merged or show a faint split as a dark thread line. 24.9. LG.

◀ The axillaries have narrower black bars than the white gaps between them. The underwing may be fairly pale and the diffuse banding of the wing coverts varies from evenly striped, as in this bird, to near-absent. See more variations in the group photo on the following page. 20.5. JL.

▶ The upperside of the wing is similar to the Great Snipe's, but mainly differs by having a wide pale trailing edge to the secondaries and a narrower one to the primaries.

The Great Snipe has only diffuse greyish-white markings on the secondaries and orderly rows of spots on the greater and median secondary coverts. 20.5. JL.

Similar species

The Common Snipe is one of four closely related species of snipes. Two of these are described separately.

Plumage and identification

Sexes are identical and show no seasonal variation. Only juveniles may be distinguished.

Overview. Blackish-brown crown with a distinct pale central stripe, wide and pale supercilium, generally dark lore. The dark line continues behind the eye and there is a more or less pronounced dark marking below each eye. The bill is long, straight and greenish-yellow, the outer third dark. Mantle and scapulars are blackish-brown with inner reddish-brown markings and white and golden outer markings running along the uppermost scapulars, creating a golden 'V' on the back. The wing coverts are paler with reddish-brown and black barring and white tips. The tail is striped, the front of the breast mottled and there is heavy mottled barring on the flanks, whereas the rest of the breast and the central belly is white. The legs are greenish-yellow.

Wings, from above and from below. See captions.

Juvenile. Distinguishable by the markings of the median coverts. See caption.

▶ An enlarged section of wing coverts from the juvenile bird below.

The light streaks at the tips of the median coverts create an undivided beige edge with only a hint of a very fine dark line, which rarely extends to the tip. 9.8. LG

Subspecies

At present, represented by two subspecies.

Subspecies *G. g. faeroeensis* breeds on Iceland, the Faeroe Isles, the Orkneys and Shetland Islands, and the nominate breeds from the remaining British Isles across Scandinavia and western Europe, east through northern and central Asia to the Bering Sea.

Voice

The call is a hoarse nasal '*krtsch*', sometimes disyllabic with the stress on the first syllable, which is heard when it takes off and occasionally from migrating groups.

The song, which is performed from the ground, is an extended, rhythmic '*bek-bek-a bek-a bek-a*'. The vibration of the stiff tail feathers during territorial flight creates a beating sound (known as 'drumming').

The Common Snipe 'sings' only with its outermost tail feathers during its song flight.

Habitat

Very common during migration, especially in autumn, near muddy fresh and brackish ponds, shallow marshes near coasts as well as river deltas and lakes.

Breeding biology

Breeds in areas with damp tussocky fields and boggy marshes. The males arrive one or two weeks before the females. The couple is monogamous though males and females mate randomly. Pairs are territorial but may breed relatively close together in suitable locations. The nest is concealed and the female incubates the eggs alone for 17–20 days. The parents divide the brood, the male caring for the first two chicks, the female the rest. The chicks fledge after 19–20 days.

The oldest known age for a ringed bird is 16 years and three months.

▲ The bird rises to an elevation of 50 metres or more and then dives at an angle of 40 degrees, allowing the air pressure on the stiff tail feathers to produce a bleating sound when the speed is between 50 and 86 kph. 9.5. JL.

▼ A more conventional song is performed from the ground or perched on a song post as here. It consists of an extended series of 'bek-a-bek-a-bek-a' notes. Notice that the tertials cover the folded primaries and secondaries and that both types of feathers end roughly at the tip of the tail. 8.5. JL.

Migration

Spring migration in Europe is from early March until early May. The birds return from western and north-western Russia and Fennoscandia from the end of June until into November. They halt at traditional moulting grounds before continuing to their winter quarters in western and southern Europe.

Small numbers may winter in northern Europe and southern Scandinavia. The British and Irish populations are partly sedentary and are joined by Icelandic birds of the subspecies *G. g. faeroeensi,* during winter.

Distribution

The European population consists of 2,670,000 to 5,060,000 pairs of breeding birds. The Russian population is the largest with 77%, followed by Iceland with 5%, Sweden with 4% and Finland with 4%. The range of the Common Snipe covers Iceland through western and eastern Europe, Russia and northern and central Asia to Kamchatka and the western Aleutians in the Bering Sea. The wintering quarters stretch from western Europe to Africa, eastwards across the Arabian Peninsula and India to southern Asia.

▲ This species is often seen over damp meadows and marshes along the coast as well as inland.

Note the distinct white trailing edge to the arm, and that the dark barring on the greyish-white underwing coverts varies from dense to nearly absent. Also, note that the bill is held slightly dipped. 18.9. JL.

▼ Adult *G. g. faeroeensis* displaying worn upper coverts, with dark lines that part the pale tips. The outer web of the scapulars is whitish and the back stripes are narrower than in the nominate. There is much discussion as to subspecies characteristics, and great plumage variation from west to east. 25.5. Iceland. HS.

WILSON'S SNIPE
GALLINAGO DELICATA

Meaning of the name
'The delicate hen-like bird'
Gallina, hen, and *ago*, like.

Presumably, the name of the genus refers to the hen-like plumage.

Delicata, delicate or delicious.

Jizz
L. 25–27 cm. Ws. 44–47 cm. Bill approx. 7 cm.

North American sister species to the Common Snipe, which it resembles closely.

A medium-sized snipe with a long, straight bill, which seems extremely long for a bird of its size.

The breast is prominent and it often appears crouched on relatively short legs. Feeds with rapid sewing machine-like movements.

In flight. The combination of an extremely long bill and the snipe jizz makes the flight silhouette easily recognisable. Flushed birds usually rise fast in zig-zag flight to considerable height and fly for a good distance before dropping into cover again.

Plumage and identification
Sexes are identical and the plumage shows no seasonal variation. Only juveniles may be distinguished.

Overview. This species was formerly regarded as a subspecies of Common Snipe. The plumage varies from dark to pale, with or without heavily streaked flanks. In the field it is probably impossible to identify birds on the ground.

In the hand, Common Snipe typically shows 14

tail feathers whereas Wilson's Snipe typically has 16.

Tail feathers in both species are orange-brown with white tips and a dark sub-terminal band.

Black bars at the centre of the orange-red feathers have been described as a decisive characteristic for Wilson's Snipe, but it is now known that they may also be seen in Common Snipe.

In flight. For the song flight at the breeding grounds Common Snipe uses only the outermost pair of tail feathers, whereas Wilson's Snipe uses the two outermost pairs.

Wings, from above. Common Snipe has a wide white trailing edge to its secondaries. Wilson's Snipe may have a narrow pale trailing edge to the wing when its plumage is fresh.

Wings, from below. The axillaries are heavily barred with wide dark bars and narrow pale bars between them. The underwing coverts are generally more evenly barred and without random pale areas where the pale bars are wider than the dark ones, as seen in Common Snipe.

Habitat and migration

Wilson's Snipe is a very rare vagrant in Europe from North America and has been sighted in the British Isles, Belgium and the Azores.

Distribution

Distributed across the North American continent from the Aleutians to Newfoundland.

Winters from central North America to northern South America.

Voice

The warning call is an abrupt, hoarse, tern-like '*krttsch*', with no stress on the first syllable as in Common Snipe.

The song, a rapid '*bick-bick-bick*', is performed perched on the ground. During territorial flight the vibrating sound, which is produced with four outer tail feathers, is fainter and hollower-sounding than that of Common Snipe.

◀ Wilson's Snipe appears nearly identical to its European sister species, the Common Snipe.
These two species are best identified by their song and call, the width of the pale trailing edge to the secondaries seen from above, and the width of dark versus pale bars on the axillaries.
23.10. Yazoo City. Mississippi. BS.

▶ The underwing is the best distinguishing characteristic. Underwing coverts are evenly barred and devoid of pale areas. The distinctive axillaries have wider dark bars than pale ones.
In Common Snipe the pale bars are wider than the dark ones. 3.10. Yazoo City, Mississippi. BS.

PIN-TAILED SNIPE
GALLINAGO STENURA

Meaning of the name
'The narrow-tailed hen-like bird'
Gallina, hen, and *ago*, like.

Presumably, the name of the genus refers to the hen-like plumage.

Greek *stenos*, narrow, *ouros*, -tailed.

Jizz
L. 25–27 cm. Ws. 44–47 cm. Bill approximately 6.5 cm.

Similar in size to Common Snipe, but has a distinctively different look with a rounder head, shorter bill, larger more Woodcock-like eyes, a rounder body and a short tail. It feeds in a calmer, more methodical fashion than Common Snipe.

In flight. Often takes wing with a short, squeaky compressed '*eerp*', with no zig-zagging. The toes extend well beyond the tip of the short tail. When the tail is worn, nearly the whole foot protrudes.

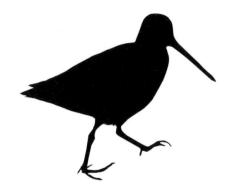

Similar species
Common Snipe and Wilson's Snipe, see these.

Plumage and identification
Sexes are identical and the plumage shows no seasonal variation. Only juveniles may be distinguished.

Overview. Compared to Common Snipe, has shorter bill, deeper at its reddish base. Also, the secondary coverts of Pin-tailed Snipe are not pale-tipped and its plumage displays less contrast than that of the Common Snipe. At close range a wide, yellowish-beige fringe on both webs of the scapulars may be seen. In Common Snipe the outer web is yellowish-beige and the inner vane reddish-brown. On perched birds the folded wings nearly extend to the tip of the short tail and the flight feathers do not protrude beyond the tertials, or only marginally so.

In flight, from above. Contrary to Common Snipe, only has a narrow pale trailing edge to the secondaries and no distinct pale tips to the wing coverts. Instead, these appear as a light greyish-brown panel, which contrasts to the rest of the dark wing.

In flight, from below. Finely barred underwing coverts and barred axillaries with slightly wider black bars than white ones.

Pin-tailed Snipe is a medium-sized, large-eyed, round-headed, short-tailed, stocky snipe.

The scapulars of this bird are very worn, the tertials nearly non-existent and the flight feathers are bleached.

This species breeds west of the Ural river in Russia and may well be overlooked among the thousands of Russian Common Snipes that migrate across western Europe in autumn. 17.9. Thailand. AJ.

▲ This species has barred underwing coverts, median and lesser wing coverts. On the barred axillaries the pale bars are as wide or wider than the dark ones. 6.3. Pak Thale. Thailand. HS.

▼ Close-up of the needle-thin outer tail feathers, only 1 mm wide near the tip. The tail feathers may be studied when the bird is preening, or during courtship display at the breeding grounds. 10.11. Israel. YK.

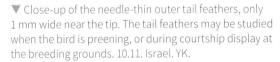

▲ In flight a pale wing panel is seen. This is made up by the greyish-brown median coverts, which lack the distinctive white-spotted tips seen in Common Snipe. Furthermore, there is no broad white trailing edge to the secondaries. 23.12. Oman. HJE.

▼ (Section) Common Snipe.
The edges of the scapulars are whitish-yellow on the outer webs and reddish-brown on the inner webs. 9.8. LG.

▼ ▼ (Section) Pin-tailed Snipe.
The scapulars have a yellowish-beige fringe to both outer and inner webs. 22.11.Oman. HJE.

Voice

The call is a creaking, short '*eerp*'.

The song, given in flight, is a cicada-like, repeated, dry, accellerating, rasping '*zhik-zhik*', followed by an angry humming sound, which is produced by the needle-thin outer tail feathers when the bird dives steeply during display flight.

Habitat and migration

Migrates over a broad front from its breeding grounds, singly or in small groups. May well be overlooked as a vagrant to western Europe. Outside of the breeding season, found in similar habitats as the Common Snipe.

Distribution

Breeds from May to August in Arctic, boreal wetland, also on scarcely vegetated damp fields and tundra up to 2,500 meters height, on the tree line.

The European population is estimated at 2,000 to 5,000 pairs, which all breed on the northern side of the Ural Mountains in northern and central Russia. Its distribution extends eastwards to north-eastern Siberia. The wintering quarters are primarily located in India and south-east Asia, but some may winter in the Middle East and Africa.

JACK SNIPE
LYMNOCRYPTES MINIMUS

Meaning of the name

'The smallest, hidden in marshland'
Greek *limne*, swampy grassland, *kruptos*, concealed.
   Latin *minimus*, the smallest.

Jizz

L. 17–19 cm. Ws. 38–42 cm. Bill approximately 4 cm. The smallest of the snipes. The size of a Dunlin, with a fairly short bill, a little over half the size of that of Common Snipe. The body appears plump and during feeding the bird bobs up and down, constantly and deeply, giving the impression that it has rubber legs. The bill is repeatedly inserted in the soft mud in the manner of a sewing machine needle. Sits tight in hiding, usually parallel to sedge and twigs so that its golden stripes make it merge into the surroundings.

In flight. Usually silent at take-off, but a faint '*gotch*' is occasionally heard. It usually flies only a short distance, calmly with none of the wild banking of Common Snipe. The wing has a white trailing edge, most obviously on the secondaries and, if well illuminated, the golden stripes of the back are conspicuous. The toes do not extend beyond the tail in flight and when alighting a dark spear-shaped tail may be seen with no white whatsoever.

Plumage and identification

The sexes are identical and there is no seasonal variation. Juveniles may only be identified in the hand by details of the undertail coverts.

Overview. Has characteristic head markings with a dark crown flanked by a wide cream supercilium, which contains a short dark 'eyebrow' stripe. Lore is dark, and cheek pale with dark bar across it. The bill is yellowish to rose at its base with a dark tip. Mantle and back have broad golden stripes and the upper scapulars have an extra pale line.

The darkest parts of the back feathers and scapulars have a metallic sheen, which alternates between dark green and blue. Neck, breast and flanks are streaked with the rest of the underside greyish-white. The legs are greyish-green to yellowish-green.

▲ The Jack Snipe sits tight, trusting its remarkable camouflage. Usually orients itself with its body running parallel to surrounding twigs and reeds so that the golden lines on its upper mandible, crown, back and wings make it hard to discern among the vegetation. 28.10. JL.

▼ All snipes seem short-necked, but when alert they stretch their medium-long necks and their appearance is transformed from clumsy to well-proportioned. 12.11. Oman. HJE.

Voice

A hoarse '*getch*' may sometimes be heard when it takes to wing. The song is a pulsating '*wakauko-wakauko*', similar to the sound of a horse galloping on a distant wooden bridge.

The song is performed perched on the ground or in song flight over its territory.

Habitat

Rarely seen and usually just one or a few individuals together. It is always shy and well concealed, both on migration and at the breeding grounds. May be seen from a hide or flushed at suitable localities such as marshes and meadows along the coast and inland.

A small number winter in northern Europe and in cold winters it may be seen near springs and open streams where the ground is not frozen, allowing it to feed.

Breeding biology

The breeding biology is largely unknown but it is presumed to be monogamous.

The nest with three to four eggs is built on isolated tussocks in quaking bogs and tundra marshes, or concealed in shrubs on drier ground. Both parents rear the brood, but it is the male that tends to the older chicks in the event of a second brood.

The oldest known age for a ringed bird is 12 years and four months.

Migration

Migration covers a wide front across Europe from March until mid May. Breeding is from mid April until early September (for Siberia). After breeding the adults moult near the breeding grounds and are unable to fly for a short period. In autumn migration follows a south-western course towards the British Isles, western Europe, the Mediterranean and south of the Sahara in Africa.

Distribution

In total, there are between 19,600 and 44,100 breeding pairs with 47% of these in Russia, 31% in Sweden and 21% in Finland. Its main distribution is from northern Scandinavia east through Russia to northern Siberia. Isolated populations breed in eastern Europe. Eastern populations winter from tropical Africa across the Arabian peninsula and Asia Minor to India and southern Asia.

▲ In flight the Jack Snipe resembles a Common Snipe, but note the bill's length and the somewhat bat-like, compressed outline. Also, the toes do not protrude beyond the tip of the tail. 31.3. HS.

▼ The Jack Snipe is a wary bird and mostly feeds on soggy ground covered in vegetation. In undisturbed areas it is occasionally seen out in the open along shore vegetation. Here it feeds in typical snipe fashion with rapid sewing machine-like movements of the bill in the soft ground, bobbing the front of the body in the manner so characteristic for this species. 14.10. HS.

The black-and-brown streaked upperparts of the Woodcock cause it to merge into the forest floor.

The bird usually sits tight and when forced to take off produces a buzzing sound from its wings. It looks heavy-bodied in flight, rapidly banking through tree trunks and shrubs.

The Eurasian Woodcock is rarely seen, but is one of the most numerous of European waders with an estimated population of 13 to 17 million individuals. 1.11. LG.

There are eight species of woodcocks in this genus. Three of these breed within the Northern Hemisphere and only one species, the Eurasian Woodcock, has its distribution within the Western Palearctic.

Woodcocks are medium-sized waders with long bills, heavy, stocky pigeon-like bodies and short tails and legs.

Woodcocks somewhat resemble the more dainty, longer-legged snipes. However, they differ in habitat, which for woodcocks is primarily swampy woodland and clearings and habits as well – woodcocks are nocturnal for most of the year.

As with snipes and the Eurasian Oystercatcher, woodcocks are among the few waders to actively feed their chicks.

EURASIAN WOODCOCK
SCOLOPAX RUSTICOLA

Meaning of the name

'The rustic-living woodcock'

Scolopax, Latin/Greek name for a woodcock.

Rusticola, The name of an unidentified game bird described by Pliny the Elder, a Roman natural scientist and military figure (23–79 A.D.)

Rusticola from *rustica*, rustic/simple and *cola*, inhabitant.

Jizz

L. 33–35 cm. Ws. 56–60 cm.

Bill approximately 7.5 cm.

A medium-sized wader with a long bill and large eyes placed high on its somewhat triangular head. The body is compact and pigeon-shaped with a short tail and legs, which are often invisible, hidden by belly feathers. Feeds mainly at night, with methodically undulating movements, probing for worms in the soft soil of forests and fields.

In flight. Sits tight, concealed on the forest floor and takes wing with a faint buzz from the wide wings. Flies rapidly following a zig-zag course, plump body hanging, avoiding tree trunks and shrubs, before dropping to the ground again. Easiest to observe over the breeding grounds in clearings at night, where males perform their territorial flight ('roding').

Plumage and identification

No seasonal variation and sexes are similar, though the male has a slightly shorter bill.

Overview. Black crown and nape with wide pale bars, large black eyes with a white lower eye-ring. The long bill is deep at its base with a faintly pink inner half, the remainder dark. The upperside is reddish-brown with black centres to the nape feathers and upper scapulars. The wing coverts are white-tipped. The pale tips of the short, orange-red tail are rarely seen in flight. The entire underside is covered in fine, mottled, blackish-brown barring with a rusty hue to the flanks. The legs are greyish to greyish-pink.

Adult breeding, hunting for earthworms, the main diet of this species. 10.6. Norway. HS.

◀ The Eurasian Woodcock is rarely seen flying in broad daylight.

It is distinguished from snipes by its wide and slightly bowed wings, its full breast and the uniformly brown underside with mottled barring. Also, snipes are slimmer, the centre of their belly is pale and they display more contrast in the barring of their underwing. 18.10. SEJ.

In flight. Upperparts are dark brown with no great contrasts. Underside shows even, greyish-brown, mottled barring.

Juvenile. In fresh plumage may be distinguished by fine mottled barring on breast and belly. Subsequently juveniles and first winter birds may only be identified in the hand by their wider, reddish-brown tips to the primary coverts, whereas those of adults have paler, narrower tips.

Habitat

The Eurasian Woodcock is very widespread and breeds in almost all European countries, where it may be observed both on migration and breeding. The species is seen most easily in the breeding territories at dusk and dawn, the male flying over swampy forest clearings in a linear display flight known as 'roding'.

Also often observed during spring and autumn migration where birds can be seen foraging in ditches and on soft ground in woodland, or flushed from hiding places.

Voice

Silent except in the breeding season, with no calls heard during migration.

The song consists of two separate elements, heard mainly at dusk but also at dawn when the male patrols a suitable breeding area in search for females.

Only heard in calm weather and at close quarters, the faintest part of the song consists of two or three croaking '*orrrw-orrrw-orrrw*' notes, followed by a chirping '*pisssip*', which most often is what alerts the observer of its presence.

Competing males or couples in pursuit over the territory emit a series of chirps: '*pli-pli-pilip-pli-pip*'.

▼ The receptive female is claimed to call the patrolling male from the forest floor but this sound has not been described. The call may be combined with a courtship dance display where the female raises her tail, flashing its white tips in the twilight.

A Eurasian Woodcock discovered roosting at day on the floor of beech forest, the soft humus under the leaves teeming with its favourite food, earthworms. 1.11. LG.

Breeding biology

At dusk and dawn the males' unusual song, consisting of growling and squeaking sounds, may be heard when they patrol suitable breeding localities, such as mixed deciduous and coniferous forest with clearings and boggy stretches.

The receptive female is thought to call the male down to her, maybe in combination with a courtship dance performance, flashing the band of white tips on the underside of the tail feathers in the twilight.

On the ground the pair performs their courtship display, stooping with erect tails and dipped bills. This culminates in mating. The male remains with the female until the eggs are laid, at which time he searches for another female in a different area. The nest is built out in the open or only slightly concealed by vegetation. The female incubates the eggs and rears the chicks alone.

The eggs hatch after 22 days and the chicks follow the female across the forest floor. Initially she feeds them, on earthworms and other small animals. The chicks are fledged after 15–22 days. There is probably a second brood.

Though rarely seen, Eurasian Woodcocks are reported by hunters and birdwatchers alike to carry chicks, either on their backs or between their legs. It is unclear whether this is normal behaviour or the chicks by chance were stuck between the legs or on the back of a female that was flushed while brooding the young.

The oldest known age for a ringed bird is 15 years and six months.

Migration

This species migrates at night only and roosts alone, but is numerous on spring migration from late February until late April, arriving at its breeding grounds between March and mid May at which time the males' territorial flight commences. The breeding period is from late February until early June. The birds remain at the breeding grounds until moult is completed. The Woodcock is a weather-dependent migrant and the autumn migration especially is influenced by weather. A sudden fall in temperature with frost in the north or north-west results in mass migration and birds

▲ Woodcock, territorial flight. Collage LG/AK.

are numerous in forests and at night on fields hunting for earthworms.

During autumn migration in October–November, birds from Fennoscandia, western Russia and the Baltic countries fly south or south-west to winter in the British Isles, central Europe and around the Mediterranean; fewer reach north Africa.

Distribution

The Woodcock is the most numerous of European waders with an estimated population of between 13,800,000 and 17,400,000 adults. It ranges from the Azores and western Atlantic isles in the west, across the British Isles, northern and central Europe and extends in a band across central Asia to Japan and the Pacific. The largest populations are in Russia with 84%, then 8% in Sweden and 2% in Finland.

Eastern populations winter primarily in Iraq, India and southern Asia.

▲ The black, grey and brown colours of the Eurasian Woodcock's plumage reflect its primary habitat – the forest floor, where it sleeps all day except in the breeding season, perfectly camouflaged. To compensate for this perilous lifestyle the Eurasian Woodcock has developed extra large, light-sensitive eyes placed well back and high on the head, giving 360-degree vision. This enables the bird to monitor its surroundings without moving its head and giving away its location. 26. 1. LG.

▼ A Eurasian Woodcock feeding with a Blackbird near a spring, in a cold spell when frost and snow makes it impossible to hunt for earthworms in the frozen forest soil.

On soft soil the Eurasian Woodcock uses the same 'paddling' technique as plovers, to lure worms to the surface. 23.12. JLA.

All three species of phalaropes breed in the Northern Hemisphere.

One of these, Wilson's Phalarope, is endemic to North America, whereas the Red-necked and Grey Phalaropes have circumpolar distribution.

Phalaropes are small to medium-sized, elegant and lively waders with lobed toes.

The lobed toes allow the bird to swim well. They spin in tight circles on the water, churning small food items to the surface where they can be snatched up.

In phalaropes the female is the more colourful, territorially aggressive and active during courtship display. When there is a surplus of males the female will mate with a new male to produce a second brood. The male incubates the eggs and rears the chicks alone.

Phalaropes migrate great distances. Outside of the breeding season, the two European species live in flocks on open sea. With their salty diet of plankton and krill, they extract and excrete excess salt from their nostrils by means of salt glands, as seen in other seabirds.

Juvenile Red-necked Phalarope.
Migration from the northern European breeding grounds covers a broad front along coastlines and inland. Juveniles may be seen in late summer along beaches with decaying seaweed, at shallow ponds with lush vegetation, coastal marshes and inland lakes. 28.8. HS.

RED-NECKED PHALAROPE
PHALAROPUS LOBATUS

Meaning of the name

'The lobed white-foot'
Greek *phalaropus*, from *phalos*, white, and *pous*, foot.
   Latin *lobatus*, lobed, eared.

Jizz

L. 18–19 cm. Ws. 31–34 cm.
A small, elegant and energetic phalarope the size of a Dunlin with a needle-fine black bill, a small head and short neck. Typically feeds while swimming, spinning with rowing, bobbing movements, while rapidly snatching flies and other food items from the surface. May also be seen feeding in brooks along the coast. Generally very tame but seems restless due to its frequent, brief excursions with abrupt landings.

In flight. Often seen skimming over the sea, rapidly banking in zig-zag flight. It may be difficult to discern among the sandpipers that it usually joins during autumn migration. The toes do not protrude beyond the tail. The safest way to identify it, when only glimpsed, is by the relatively short and very slender, pointed bill.

Similar species

At all ages may be confused with Grey Phalarope of the same age.

Important differences are the black, slender, pointed bill and the grey primary underwing coverts of the Red-necked Phalarope. Grey Phalarope has a stouter bill, often with a yellow base in winter and pure white undersides to the wings. Furthermore, the Grey Phalarope in flight appears less dainty, heavier with longer wings.

Note that this species is often seen with Calidris sandpipers, from which it may be distinguished by the shape of its bill and its distinct wide, white wing-bar.

Plumage and identification

Adult breeding, female. Blackish-grey head, neck and sides of the breast with white throat and short, narrow white supercilium immediately above the eye. Along the sides of the neck a distinctive orange-red band that extends to beneath the white throat. The upperparts are dark grey and have two parallel golden stripes across mantle and back, supplemented by golden webs to the uppermost scapulars. The sides of the breast are grey, the flanks diffusely greyish and the rest of the underside is white. The greyish-black legs have partially webbed toes.

In flight, from above. The upperparts are uniformly grey with golden stripes on the back and a distinct, wide white wing-bar extending from the body along the secondary coverts to the middle of the primary coverts. The rump is white with a dark stripe down the middle and the tail has grey central feathers with white outer webs to the outer tail feathers.

In flight, from below. The underwing displays three colours, dominated by white secondary coverts contrasting with the medium grey

Adult breeding, male. The male is paler than the female with more and wider golden-tan fringes to the scapulars. 23.5. Iceland. JL

primaries and secondaries and black small coverts along the wing's leading edge.

The Grey Phalarope's underwing is mainly white.

Adult non-breeding. Grey-and-white head with a black bar behind each eye and varying amount of black at the back of the crown. The upperparts are light grey with white feather fringes and a white-striped mantle. The underside is white.

Juvenile. Similar to adult in non-breeding plumage, but the head is darker with more black on the crown over the pale forehead and a wider bar behind the eye. Also, blackish-brown upperparts with golden, eventually white, back stripes and golden to tan fringes to the upper scapulars. The breast is sometimes diffusely beige contrasting with a white underside, with flanks and sides of the breast mottled in grey. The legs are greyish-yellow to pale pink.

First winter. Distinguished from adult non-breeding birds by retaining some yellowish-white fringes to the scapulars and tertials well into spring.

Voice

The call is a faint, hard '*bik*' and a finer, sputtering, agitated, repeated '*prret*' which is heard during courtship display and mating, sometimes in longer phrases. This call is also heard from feeding flocks in open sea.

Habitat

The Red-necked Phalarope is a scarce migrant. Most are sighted near shallow coastlines with rotting seaweed and in ponds and lakes along the coast. Seen singly or in groups of a few birds, often together with stints.

In its winter quarters on the open sea, however, it gathers in large flocks made up of hundreds of birds.

▲ Adult breeding, female. Red-necked and Grey Phalaropes are difficult to distinguish in winter. However, the Red-necked Phalarope has a black, white and grey underwing, whereas the Grey Phalarope's underwing coverts are white with a touch of black at the carpal joint and a grey wash to the trailing edge of the wing. 21.5. Iceland. JL.

▼ Adult breeding. Mating male and female. The female has richer coloration compared to the male: notice the black streak under the male's eye. In winter plumage in adults and juveniles alike, this becomes a black bar behind the eye. 23.5. Iceland. JL.

Breeding biology

Breeds around ponds, bogs, small lakes and rivers. Breeds at one year of age, in individual pairs or in scattered colonies inland or along the coastline. The pair is monogamous and the female takes the lead during courtship display and defence of the territory. She may have a second brood with a different male. After laying her eggs she leaves the scantily lined nest, usually near the water's edge, to the male. He incubates the eggs, which hatch after 17–21 days and raises the brood alone. The chicks are independent after a fortnight and fledged after about 20 days. The female leaves the breeding grounds after the eggs are hatched, the male two weeks later. The oldest known ringed bird was nine years and 11 months old.

Migration

Spring migration to the Fennoscandian and Russian breeding grounds, individually and in small groups, covers a wide front, primarily across eastern Europe. The birds arrive at their breeding grounds from late April until mid June.

Adult females leave the breeding grounds from late June, followed by adult males from late July and finally immature birds in August and September.

Data from geologgers mounted on birds from the Shetlands suggest that western European birds from the British Isles and possibly the north Atlantic isles winter at sea, off the coast of Ecuador and Peru. Their route went across the northern Atlantic to Labrador and continued along the coastline of North and Central America in both directions. During autumn migration few are sighted in western Europe since Fennoscandian and Russian breeders travel south-eastwards over land to their moulting grounds, the Black Sea and the Caspian Sea, before moving on to their winter quarters in the Arabian Sea where they gather in large flocks far from land.

Eastern Russian populations winter at sea in south-east Asia.

▲ Juvenile is distinguished from juvenile Grey Phalarope by its finer bill, slimmer body, narrower wing-bars and golden stripes on its back, which are retained well into autumn. Juvenile Grey Phalarope has similar back stripes but these are already moulted in late summer and replaced by the grey first winter plumage. 28.8. HS.

▼ Adult breeding, male at the nest. After mating, while the male incubates the eggs, it is the female that defends the territory. But briefly after the eggs are hatched she leaves the family, sometimes to produce a second brood with a different male. 29.6. Norway. HS.

Distribution

The European population is estimated at between 550,000 and 1,200,000 adult birds. Of these, 74% breed in Russia, 12% in Iceland, 4% in Sweden, 2% in Norway and 2% in Finland.

Breeds on one or two Scottish islands, the north Atlantic isles, central and northern Fennoscandia and along all other Arctic coastlines and in bogs and boreal tundra.

▶ Juvenile.
The slender, black, needle-like bill distinguishes this species from Grey Phalarope around the year. From early autumn, golden stripes on the back distinguish juveniles from juvenile Red Phalarope.
Furthermore, the Grey Phalarope's bill is yellowish at the base, its neck has a rose hue, grey winter plumage feathers on its back and shoulder, and the tertials have pinkish-beige fringes, not golden as in juvenile Red-necked Phalarope. 1.9. LG.

▶ Adult moulting to non-breeding plumage.
Aged by the absence of golden webs to scapulars and tertials and by bleached, worn tail feathers, flight feathers and tertials contrasting to the fresh greyish-blue tail feathers and tertials with narrow pale edges. 10.11. Oman. ISA.

▶ Adult non-breeding.
Aged by nearly white head, greyish-blue upperparts with wide white fringes to the scapulars and the fresh tertials with narrow white edges; additionally, by a few worn flight feathers, barely visible over the white, fringed undertail coverts.
Notice the white-streaked mantle, which in winter is greyish-blue in Grey Phalarope, and the Red-necked Phalarope's slender, more delicate appearance with a slender throat and a flat back; at a distance this distinguishes it from the stockier Grey Phalarope with its more curved back. 3.10. China. DP.

GREY PHALAROPE
PHALAROPUS FULICARIUS

Meaning of the name

'The coot-like white-foot'
Greek *phalaropus*, *phalos*, white, and, *pous*, foot.
Latin *fulicarius*, *fulica*, coot.
The resemblence of its lobed feet to those of the coots inspired the name.

Jizz

L. 20–22 cm. Ws. 37–40 cm.
A stocky phalarope with a plump, pigeon-like body and a short, thick neck. The bill is thick and blunt-tipped and the legs are short.
When feeding, spins and turns, whirling up food items, which are snatched from the surface. Also hunts for insects along the shore.

During migration, occasionally birds are accidentally blown towards the shore where they may be seen flying low over the surf in search for food, or fearlessly plunging into waves for prey.

In flight. Longer and wider wings than the Red-necked Phalarope and appears more rotund and heavy in flight due to its rounded back and full breast. The toes do not extend beyond the tail in flight. Often very tame.

Similar species

During autumn and early winter, birds blown ashore may be confused with Sanderlings, but are identified by the black bar behind each eye, shorter white wing-bars and stockier, pot-bellied appearance, as well as by its fearless feeding habits in surf and waves. At a distance, may resemble a small, plump, greyish-white gull.

Plumage and identification

Adult breeding, female. Black crown, white around eye, and a dark-tipped short, thick and straight bill. Mantle, scapulars and back feathers are conspicuously black at the centre with yellowish-white fringes. The neck and underside is brick red. The short legs have partially webbed, lobed toes and are greyish-yellow to greyish-pink.

Male. Similar to the female but the crown is blackish-brown with orange streaks, less white on the face, and golden fringes to scapulars and back feathers. The underside is paler red and appears more subdued and mottled than on the female.

In flight, from above. The wings are grey with wide white wing-bars extending from the body to the median wing coverts. The back is dark with golden stripes. Upper tail coverts are brick red and contrast with the grey tail with its dark central feathers.

In flight, from below. Mainly white underwing coverts with only little greyish-black on the lesser and median primary underwing coverts. See Red-necked Phalarope, which has more contrast to its underwing.

Adult breeding. A breeding pair. Male to the left, female to the right. 15.6. Iceland. HJE.

Adult non-breeding. Resembles a Red-necked Phalarope with a whitish head, black at the back of its crown, a black bar behind each eye, greyish-blue upperparts and white underside.

Distinguished from Red-necked Phalarope by its uniform mantle, devoid of white stripes, its powerful yellow-based black bill and its stocky body with curved back.

Juvenile. Juveniles in fresh plumage are rarely seen outside of the breeding area. They have a greyish-black bill, black crown and a short white supercilium. The greyish-black neck contrasts to the blackish-brown upperparts with its golden fringes to mantle, scapulars, tertials and tail feathers. Wing coverts are greyish with pale fringes. Throat, upper breast and flanks are greyish-pink to salmon and the remaining underside is white. The legs are greyish-pink.

Altogether, gives the impression of a very dark bird with golden back stripes.

Post juvenile. Black crown and nape parted by a white supercilium. The remainder of the head and neck is beige-pink to faint salmon. Back feathers and scapulars are uniformly greyish blue with some remaining black and golden; the tertials are black with narrow, beige-pink edges. Upper breast and flanks are greyish-blue, the rest of the underside white.

First winter. Unlike post juvenile, has a yellowish base to its bill, a white head, black at the back of its nape, a black bar behind its eye and uniform bluish-grey mantle and scapulars.

Habitat

A scarce migrant, primarily seen in autumn when storms blow birds inshore and occasionally further inland where it may be found on reservoirs, lakes or large rivers. At the seaside migrating birds are either seen in flight low over the waves, or feeding in various ways; swimming, in fluttering flight snatching

Adult breeding. Female on left, male to the right. The female has richer colours with stronger contrast between its entirely black crown, pure white face patch, blacker upperparts and rich brick-red underside. The male shows golden 'tiger-striped' upperparts. 22.6. Iceland. MV.

food items, or running along the beach, sometimes fearlessly plunging into the surf.

Voice

The flight call, which is heard all year round, is a brief, hard '*pit*'. At the breeding grounds this call is heard as well as a more extended '*wheiiip*', possibly an alarm call.

The song is a warbling, stint-like '*pe-reep pe-reep*', performed by the female during courtship flight, which is intended to attract males.

Breeding

Breeds from one year old on stretches of tundra along the coastline with ponds and small lakes, as well as rivers and river complexes. At the breeding grounds, females perform courtship display flights in an attempt to attract unpaired males; the males pick their choice of mate.

The pair is partially monogamous, but given a surplus of males the female will leave her first mate to produce a second brood with a different male.

The nest with its three to four eggs is built concealed in grass near water, usually in colonies of Arctic Terns which help drive away predatory gulls.

The male incubates the eggs, which hatch after 18–20 days and tends to the chicks until they are fledged after 16–20 days. Juveniles gather in larger groups in lakes in the vicinity.

Migration

The Grey Phalarope migrates to its winter quarters mainly at sea far from land. Consequently, it is a rare vagrant in Europe from the north Atlantic and Arctic regions. The few annual sightings are mostly from August to November, peaking from October to mid November.

Known winter quarters, where birds gather in large flocks, are off the coast of west Africa, the western Atlantic off the shores of Central and southern North America and off the Pacific coasts of southern North America and South America.

There is some uncertainty as to there are wintering quarters in the south-western Atlantic and the Arabian Sea. That birds may winter in the latter is indicated by sightings in a number of countries in central and eastern Europe, including Bosnia-

▲ Post juvenile Grey Phalarope is distinguished from juvenile Red-necked Phalarope by its thick bill, slightly wider wing-bar, grey back with no golden stripes and pot-bellied appearance. 27.9. SEJ.

Herzegovina, Bulgaria, Croatia, Greece, Hungary, Kazakhstan, Montenegro, Poland, Romania, Serbia, Slovakia, Slovenia and Ukraine.

Distribution

The European breeding population consists of 640 to 2,400 adult birds, which primarily breed in Svalbard (Norway), with a smaller population in Iceland and a few pairs in European Russia.

The species has a circumpolar distribution across Siberia and northern North America with an estimated world population of between 1,100,000 and 2,000,000 individuals.

▶ Post juvenile in moult to first winter plumage. Aged by remnants of the blackish-brown juvenile head markings, the fresh grey mantle with remnants of beige-fringed black feathers and the distinctive juvenile tertials with whitish-beige edges.

Notice the difference between this September bird and the October bird below right, which has a more juvenile appearance to its head and neck. The first may be from a first brood, the latter probably from a second one. 20.9. HS.

▶ First winter.
Identified by juvenile blackish-brown tertials with narrow white or pinkish-beige edges. Adult non-breeding birds' tertials are grey as the rest of the upperparts.

Furthermore, the Grey Phalarope may always be distinguished by its yellow-based, fairly robust bill and in winter by its uniform greyish-blue mantle devoid of whitish streaks. 11.12. NLJ.

◀ Post juvenile with remnants of a pinkish-beige to salmon hue to its head, neck and underside and the remains of a black mantle with juvenile, golden streaks. The crown is in moult to white and the back is becoming increasingly bluish-grey. The black tertials display telltale narrow pinkish-rose edges.

The Grey Phalarope winters at sea, but during storms is blown ashore along western European coastlines where it may be observed hunting in the surf for amphipods and other prey. 15.10. EFH.

WILSON'S PHALAROPE
STEGANOPUS TRICOLOR

Meaning of the name
'The three-coloured web-foot'
Greek *steganopous*, web-footed.

From *stego*, to cover and *pous*, foot.

Latin *tricolor*, three-coloured.

This species is often placed in *Phalaropus*, but given that it does not have pale legs and feet, *Steganopus* makes more sense.

Jizz
L. 22–24 cm. Ws. 35–38 cm.

Reminiscent of a slightly larger version of Red-necked Phalarope, with a longer, black bill and longer legs.

It is the longest-necked of the three species and its appearance and habits are more like those of a sandpiper.

Adult breeding, male. As in the European phalaropes, the male is less colourful with fainter markings than the female. Note that this species never displays any white wing-bar on the uniform upperparts. 25.6. British Columbia. KK.

Like the two other phalarope species, it feeds by churning up food items from the sea bed while swimming in circles, but it also wades more frequently than the other species, skimming off food items from the surface with rapid sideways movements of the bill like an avocet. Hunts insects and other small creatures on mudflats and lakeshores, body inclined, often with bobbing, mechanical movements.

In flight. The wings are longer and wider than in the two European species. In flight the toes protrude beyond the tail.

Plumage and identification
Adult breeding, female. Grey crown and white supercilium. Black lore and a black eye-stripe that extends to the nape and down the back of the neck, where it merges into the chestnut of the sides of the neck, creating a contrast to the grey mantle and rusty neck.

Chin and cheeks are white and the needle-like bill is black. Scapulars are grey and chestnut, creating a diffuse contrast to greyish-brown wing coverts and tertials. Breast and belly are white and legs are black.

Male. Fainter markings than the female with uniform brown upperparts.

In flight, from above. Uniform greyish-brown upperparts with none of the female's chestnut streaks on mantle and shoulder; rump is whitish and tail feathers greyish-brown with whitish inner vanes. Very narrow, nearly non-existent pale trailing edge to primary and secondary coverts, which distinguishes it from the other two phalaropes, which both display wide white wing-bars.

In flight, from below. Greyish-white underside with little contrast, and no black on the lesser primary coverts.

Adult non-breeding. Pale bluish-grey upperparts and white underside. Distinguished from the other species by its fairly long, needle-like bill, elongated body shape, grey crown, white supercilium and a narrow, less distinct black stripe behind each eye.

Juvenile. See caption.

First winter. See caption.

Voice

Mostly silent except in the breeding season.

The flight call is a short, deep, mellow, duck-like '*woeep*'.

Adult breeding, female. In breeding plumage, the beautifully marked female is easily distinguished from all other waders. 4.6. Canada. DP.

Habitat and migration

Rare vagrant from North America to Europe. Sightings from Belgium, Bulgaria, Czech Republic, Denmark, Estonia; Finland, France; Germany, Greece, Iceland, Ireland, Italy, Latvia, Netherlands, Norway, Portugal, Spain, Sweden and the United Kingdom.

Migration from North America to its wintering quarters in western and southern South America is from late June to early November; the birds return from March to May.

Does not winter at sea as the European phalaropes, but rather around lagoons, deltas, mudflats and saline lakes.

This species is endemic to North America, its main distribution being the western part of Canada, the northern part of western USA and around the Great Lakes.

Distribution

Breeds on prairie as well as on farmed grassland with ponds and at the edges of grassy bogs up to the taiga zone.

Post juvenile with a few fresh, grey mantle feathers and scapulars. Easily distinguished from adults and juveniles of other phalaropes by its yellow, medium-length legs. Other field characters are the greyish-brown upperparts with a faint 'V', and its whitish-beige underside. 25.8. DP.

▲ Adult breeding to non-breeding.
A swimming Wilson's Phalarope is reminiscent of a more elongated Red-necked Phalarope and may be distinguished from this in all plumages by its longer bill. 7.7. Salinas, California. BS.

▼ First winter.
Identified by its grey crown, white supercilium, narrow black eye-stripe, long neck and elongated body shape.
Its age is determined by juvenile, black tertials with narrow white edges. 28.9. RSN.

▲ Juvenile.
Contrary to European phalaropes, this species has uniform upperparts, white rump and at the most a very narrow white wing-bar along its greater wing coverts. Additionally, juveniles have white-rimmed median secondary coverts. 12.8. California. DP.

Main contributors of photos (numbers in parenthesis)

DP – Daniel Pettersson
www.danielpettersson.com (84)
HJE – Hanne og Jens Eriksen
www.birdsoman.com (31)
HS – Helge Sørensen
www.birdphotos.dk (89)
JL – John Larsen
KBJ – Klaus Bjerre
www.kbphoto.dk (24)
KK – Kevin Karlson
www.kevintkarlson.com (36)
LG – Lars Gejl
www.larsgejlfoto.dk (119)
NLJ – Nis Lundmark Jensen (34)

Other contributors of photos

AA – Aurelién Audevard
www.oiseaux.net (6)
AK – Axel Kielland (4)
AJ – Ayuvat Jearwattakananok (2)
BLC – Bo L. Christiansen (13)
BS – Brian Sullivan (6)
EFH – Eva Foss Henriksen (11)
GV – Gerrit Vyn (5)
HHL – Hans Henrik Larsen (1)
ISA – Ib Steen Andersen (7)

JH – Julian Hough (1)
JKAM – Jan Van den Kam (7)
JLA – Johnny Lauersen (1)
JP – Jari Peltomäki (1)
JSH – Jens Søgaard Hansen (1)
KF – Keith Fox (1)
LK – Lior Kislev (8)
MV – Markus Varesvuo (5)
MYJ – Manjeet og Yograj Jadea (1)
PN – Peter Nielsen (1)
RSN – Rune Sø Nergaard (1)
SD – Sergey Dereliev
www.dereliev-photography.com (3)
SKKO – Silas K.K. Olofson (1)
SEJ – Steen E. Jensen (3)
SP – Stefan Pfützke (1)
SSL – Stephan Skarup Lund (1)
TL – Tomas Lundquist (2)
TO – Torben Olsen (1)
TVN – Thomas Varto Nielsen (1)
TH – Tommy Holmgren (7)
WRT – Warvick Tarbotton (1)
YK – Yosef Kiat (1)

We have gone to great lengths to locate and credit all contributors of photos and bird voices. Should any be omitted, we apologise and will correct this in later editions.

REFERENCES

Literature and periodicals

Chandler, Richard. Shorebirds of North America, Europe and Asia. Princeton 2009.

Gejl, Lars. Kend Fuglen. Gyldendal 2012.

Gregersen, Jens. Arktisk Sommer. Gyldendal 2014.

Jobling, James A. Helm Dictionary of Scientific Bird Names. Published by Christopher Helm. London. 2010.

Kam, Jan Van De. Bruno Ens. Theunis Piersma. Leo Swarts. Shorebirds. An illustrated behavioural ecology. KNNV Publishers. 2004.

Lange, Peter (red) Fugleåret 2013. Dansk Ornitologisk Forening.

Lauersen, Karsten. John Frikke. Rastende vandfugle i Vadehavet. DOF tidsskrift 107. Nr. 1. 2013.

Lederer, Roger. Carol Burr. Latin for Birdwatchers. Quid Publishing. 2014.

Meltofte, Hans. Jon Fjeldså. Fuglene i Danmark. Gyldendal 2002.

Message, Stephen. Don Taylor. Shorebirds of North America, Europe, and Asia. A guide to field identification. Princeton University Press. 2005.

Nyegaard, Timme. Hans Meltofte. Jesper Toft. Michael Borch Grell. Truede og sjældne ynglefugle i Danmark, 1998-2012. Dansk Ornitologisk Forenings Tidsskrift 108. Nr. 1. 2014.

O´Brien, Michael. Richard Crossley. Kevin Karlson. The Shorebird Guide. Houghton-Mifflin Compagny. 2006.

Staarup Christensen, Jørgen. Palle Ambech Frænde Rasmussen. Revideret status for sjældne fugle i Danmark før 1965. Dansk Ornitologisk Tidsskrift 109. Nr. 2. 2015.

Svensson, Lars. Killian Mullarney, Dan Zetterström. Fågelguiden. Europas och Medelhavsområdets fåglar i fält. Bonnier Fakta 2009.

Online articles and links

A toolkit for finding Slender-billed Curlews
http://www.rspb.org.uk/Images/sbcleaflet_tcm9-203901.pdf

Advances in the Field Identification of North American Dowitchers. Cin-Ty Lee. Andrew Birch.
http://static1.squarespace.com/static/54b9bb6fe4b07b4a7d145b55/t/54edd57be4b0978eebcea523/1424872827140/2006LeeBirchDowitcher.pdf

Amerikansk præstekrave/Semipalmated plover
http://bna.birds.cornell.edu/bna/species/444/articles/characteristics

Bill length and bill shape of Semipalmated Sandpiper
http://britishbirds.co.uk/wp-content/uploads/article_files/V89/V89_N05/V89_N05_P234_236_N054.pdf

Black-tailed Godwits Sub-Specific Identification & Status in the County.
Chris G. Knox.
http://www.ntbc.org.uk/wp-content/uploads/2013/06/Black-tailed-Godwits.pdf

Displaying Swinhoe's Snipe in eastern European Russia: a new species for Europe. Vladimir V. Morozov
http://britishbirds.co.uk/wp-content/uploads/article_files/V97/V97_N03/V97_N03_P134_138_A003.pdf

Field characters for ageing and sexing Stone-curlews
RE Green, CGR Bowden – British Birds, 1986 – britishbirds.co.uk
http://britishbirds.co.uk/wp-content/uploads/article_files/V79/V79_N09/V79_N09_P419_422_A101.pdf

Field identification of Pintail Snipe S. C. Madge
http://britishbirds.co.uk/wp-content/uploads/article_files/V70/V70_N04/V70_N04_P146_152_A037.pdf

Geographical segregation in wintering Dunlin Calidris alpina populations along the East Atlantic Flyway: evidence from mitochondrial DNA analysis.
Ricardo J. Lopes. Liv Wennerberg.
http://jncc.defra.gov.uk/PDF/pub07_waterbirds_part4.4.11.pdf

Handbook of the Birds of the World Alive. Lynx Edicions, Barcelona.
Jobling, J. A. (2015). Key to Scientific Names in Ornithology. In: del Hoyo, J., Elliott, A., Sargatal, J., Christie, D.A. & de Juana, E. (eds.) (2015).
http://www.hbw.com/

Incubation period and foraging technique in shorebirds
http://www.jstor.org/stable/2461289

Identification, taxonomy and distribution of Greater and Lesser Sand Plovers.
Erik Hirschfeld, C. S. (Kees) Roselaar and Hadoram Shirihai.
http://www.californiabirds.org/members/sand-plover.pdf

In the world of Ruff´s, a male bird that´s sneaky and well endowed.
By Leslie Evan Ogden. April 2014.
http://www.earthtouchnews.com/natural-world/animal-behaviour/in-the-world-of-ruffs-a-male-bird-thats-sneaky-and-well-endowed

Migratory Connectivity of Semipalmated Sandpipers: Winter Distribution and Migration Routes of Breeding Populations.
Cheri Gratto-Trevor, R. I. Guy Morrison, David Mizrahi, David B. Lank, Peter Hicklin and Arie L. Spaans http://www.sfu.ca/biology/wildberg/papers/Grattoetal2012Waterbirds.Connectivity.pdf

Northumberland & Tyneside Bird Club
Chris. G. Knox. Blacktailed Godwits Sub-Specific Identification & Status in the County.
http://www.ntbc.org.uk/wp-content/uploads/2013/06/Black-tailed-Godwits.pdf
notes on the biometrics and egg measurements of breeding dunlins in sutherland, scotland by John Barrett and Catrina Barrett.
https://sora.unm.edu/sites/default/files/journals/iwsgb/n046/p00029-p00030.pdf

Population development of southern Dunlin. Christof Herrmann, Ole Thorup.
http://helcom.fi/baltic-sea-trends/environment-fact-sheets/biodiversity/population-development-of-southern-dunlin

Report to OSME, RSPB and Viking Optical Limited on the Spring 2008 Curlew Numenius survey of Lake Ayaqaghitma, Uzbekistan.
http://www.rspb.org.uk/Images/tripreportaug08_tcm9-203233.pdf

Ruff projekt. David B. Lang.
http://www.sfu.ca/biology/wildberg/ruff.html

Social behavior of the Ruff. A. J. Hogan-Warburg.
https://books.google.dk/books?id=EDcVAAAAIAAJ&pg=PA155&lpg=PA155&dq=different+males+pugnax&source=bl&ots=pP2eIGvz9Z&sig=Nh7G0NPhE3yUTamHqTVw1kyl7jE&hl=da&sa=X&ei=N6XxVM7xE4HjONuZgYAJ&ved=0CG4Q6AEwDQ#v=onepage&q=different%20males%20pugnax&f=false

Special IWC Newsletter focuses on the quest to find the Slender-billed Curlew
http://www.unep-aewa.org/en/news/special-iwc-newsletter-focuses-quest-find-slender-billed-curlew

Surfbird.com
http://www.surfbirds.com/ID%20Articles/dowitchers1005/dowitchers.html

The functional morphology of male courtship displays in the Pectoral Sandpiper (Calidris melanotos) Tobias Riede1*, Wolfgang Forstmeier2, Bart Kempenaers2, and Franz Goller3
http://www.bioone.org/doi/abs/10.1642/AUK-14-25.1

The Royal Society for the Protection of Birds (RSPB)
http://www.rspb.org.uk/whatwedo/projects/details/198450-slender-billed-curlew#downloads

The Wilson Bulletin.
On aerial and ground displays of the world´s snipes. George Miksch Sutton.
https://sora.unm.edu/sites/default/files/journals/wilson/v093n04/p0457-p0477.pdf

You snooze, you lose. Less sleep leads to more offspring in male pectoral sandpipers.
http://www.mpg.de/5976385/sleeplessness_sandpipers

Migration and Ringing

Sortgrå ryle, Lars Hansen.
http://pandion.dof.dk/artikel/sprog%C3%B8-forbindelsen-til-det-h%C3%B8je-nord

Websites

Birdlife International.
http://www.birdlife.org/

Dansk Ornitologisk Forening
http://www.dof.dk/

Global Flyway Network
http://globalflywaynetwork.com.au/

HBW. Handbook of the Birds of the World.
http://www.hbw.com/

International Wader Study Group
http://www.waderstudygroup.org/about-us/

Netfugl
http://www.netfugl.dk/

The IUCN Red List of Threatened Species.
http://www.iucnredlist.org/

The Royal Society for the Protection of Birds (RSPB)
http://www.rspb.org.uk/

Wetlands International
http://www.wetlands.org/

Wildscreen Arkive
http://www.arkive.org/

Fuglestemmer/DVD
Xeno-canto
http://www.xeno-canto.org/explore?query=

Macaulay Library
http://macaulaylibrary.org/

Waders. Paul Doherty. DVD.

INDEX